Introduction to
BIOSTATISTICS

Introduction to
BIOSTATISTICS

A. D Leguina
Juan Pascal

Kruger Brentt
Publishers
2025

Kruger Brentt Publishers UK. LTD.
Company Number 9728962

Regd. Office: 68 St Margarets Road, Edgware, Middlesex HA8 9UU

ISBN: 978-1-78715-440-7

Disclaimer:

Every possible effort has been made to ensure that the information contained in this book is accurate at the time of going to press, and the publisher and author cannot accept responsibility for any errors or omissions, however caused. No responsibility for loss or damage occasioned to any person acting, or refraining from action, as a result of the material in this publication can be accepted by the editor, the publisher or the author. The Publisher is not associated with any product or vendor mentioned in the book. The contents of this work are intended to further general scientific research, understanding and discussion only. Readers should consult with a specialist where appropriate.

Every effort has been made to trace the owners of copyright material used in this book, if any. The author and the publisher will be grateful for any omission brought to their notice for acknowledgement in the future editions of the book.

For information on all our publications visit our website at http://krugerbrentt.com/

PREFACE

Biostatistics and research methods are integral components of the fields of public health, medicine, and epidemiology, serving as the backbone for evidence-based decision-making and scientific inquiry. Biostatistics involves the application of statistical principles and methodologies to analyze data related to health and biological sciences. It provides tools for designing studies, collecting and analyzing data, and interpreting results to draw valid conclusions about health-related issues.

Researchers utilize bio-statistical techniques to understand the patterns of diseases, evaluate the effectiveness of treatments, and identify risk factors associated with health outcomes. For instance, biostatistics plays a crucial role in clinical trials, where it helps determine sample sizes, randomization processes, and data analysis methods, ensuring that the findings are both reliable and generalizable. Similarly, in observational studies, biostatistics aids in controlling for confounding variables, enabling researchers to assess associations between exposures and outcomes accurately.

Research methods, on the other hand, encompass the broader spectrum of techniques and approaches employed to conduct research effectively. These methods can be qualitative, quantitative, or mixed methods, each providing unique insights into health phenomena. Quantitative research methods focus on numerical data and statistical analysis to test hypotheses, while qualitative methods explore complex behaviors, perceptions, and experiences through interviews, focus groups, and ethnographic studies.

The integration of both methods, known as mixed methods research, allows for a more comprehensive understanding of health issues by combining statistical rigor with rich, contextual insights. The design of research studies is a critical aspect of both biostatistics and research methods, influencing the validity and reliability of the findings. Proper study design involves selecting the appropriate methodology, defining the population of interest, and establishing clear objectives and hypotheses. Randomized controlled trials (RCTs) are considered the gold standard in clinical research due to their ability to minimize bias and establish causal relationships.

However, observational studies, such as cohort and case-control studies, are also essential, especially when RCTs are impractical or unethical. Ethical considerations are paramount in research, ensuring that the rights and welfare of participants are protected. Researchers must adhere to ethical guidelines, obtain informed consent, and ensure confidentiality and data security throughout the research process. Moreover, the role of biostatistics extends to the interpretation and communication of research findings. Clear and effective communication is vital for translating complex statistical results into actionable insights for policymakers, practitioners, and the general public. Understanding statistical concepts such as p-values, confidence intervals, and effect sizes is crucial for evaluating the significance and relevance of research findings.

Additionally, biostatistics contributes to systematic reviews and meta-analyses, where data from multiple studies are aggregated to derive broader conclusions, enhancing the robustness of evidence in healthcare decision-making.

Biostatistics and research methods are essential for advancing knowledge in the health sciences, enabling researchers to explore intricate relationships between variables, assess interventions, and inform public health policies. The continuous evolution of statistical techniques and research methodologies ensures that researchers can address emerging health challenges effectively, fostering a deeper understanding of health disparities, disease prevention, and health promotion. As the fields of biostatistics and research methods advance, they will continue to play a pivotal role in shaping the future of healthcare, ultimately contributing to improved health outcomes for populations worldwide.

A. D Leguina

Juan Pascal

CONTENTS

CHAPTER 1

SCOPE OF STATISTICAL METHODS IN MEDICINE

Statistical methods have long been integral to medical research and practice, playing a critical role in the development, evaluation, and application of new treatments and interventions. Medicine, being an evidence-based discipline, relies heavily on quantitative data to make informed decisions about patient care, disease prevention, and treatment effectiveness. The scope of statistical methods in medicine extends across multiple domains, from designing clinical trials and analyzing patient data to improving healthcare policies and identifying public health trends.

1.1 STATISTICAL METHODS IN CLINICAL RESEARCH

Clinical research is one of the most prominent areas where statistical methods are applied. Medical researchers use statistical techniques to design, conduct, and analyze clinical trials, which are studies conducted to evaluate the safety and efficacy of new treatments or interventions. A well-designed clinical trial requires careful planning of sample size, randomization, and the selection of appropriate statistical tests to ensure that the results are valid and reliable.

1.1.1 Sample Size Determination

Before a clinical trial begins, determining the appropriate sample size is critical. The sample size impacts the study's ability to detect meaningful differences between treatment groups. Too small a sample may lead to inconclusive results, while too large a sample could be unnecessarily expensive and time-consuming. Statistical methods, such as power analysis, help researchers determine the optimal sample size required to achieve the desired level of statistical significance while minimizing errors.

1.1.2 Randomization and Blinding

Randomization is a fundamental principle in clinical trials that helps eliminate bias and ensures that patients are randomly assigned to treatment groups. By using statistical techniques to generate random allocation sequences, researchers can ensure that the trial's results are not influenced by selection bias. Blinding, another important technique, involves concealing the treatment allocation from participants and researchers to avoid bias in the interpretation of results. Statistical methods are used to design and implement both randomization and blinding strategies.

1.1.3 Hypothesis Testing and Confidence Intervals

In clinical research, statistical hypothesis testing is used to determine whether there is sufficient evidence to support a specific claim about a treatment's efficacy. For example, a researcher may test whether a new drug significantly reduces symptoms compared to a placebo. Statistical tests, such as t-tests or chi-square tests, are employed to analyze the data and draw conclusions. Additionally, confidence intervals provide a range of values within which the true effect size is likely to lie, offering a measure of the precision of the results.

1.2 STATISTICAL METHODS IN EPIDEMIOLOGY

Epidemiology, the study of disease distribution and determinants in populations, heavily relies on statistical methods to identify patterns and causes of diseases. By analyzing data from large populations, epidemiologists can uncover trends, risk factors, and the impact of interventions on public health.

1.2.1 Descriptive Statistics

Descriptive statistics are used to summarize and describe the basic features of data collected in epidemiological studies. These statistics include measures of central tendency (such as the mean, median, and mode) and measures of variability (such as standard deviation and interquartile range). By providing a clear summary of the data, descriptive statistics help epidemiologists understand the overall picture of health outcomes within a population.

1.2.2 Regression Analysis

Regression analysis is a powerful statistical tool used in epidemiology to examine the relationships between risk factors and health outcomes. For instance, logistic regression can be used to analyze the relationship between smoking and lung cancer, while linear regression can be applied to study the effect of cholesterol levels on heart disease risk. These models allow researchers to adjust for confounding variables and assess the independent effect of each factor on the health outcome of interest.

1.2.3 Survival Analysis

In medical research, survival analysis is often used to analyze time-to-event data, such as the time from diagnosis to death or the time until the recurrence of a disease. Techniques like the Kaplan-Meier estimator and Cox proportional hazards models are commonly employed to estimate survival probabilities and compare survival rates between different treatment groups. Survival analysis provides valuable insights into the effectiveness of treatments and the progression of diseases over time.

1.3 STATISTICAL METHODS IN DIAGNOSTIC MEDICINE

Diagnostic accuracy is crucial in medicine, as it directly impacts patient outcomes. Statistical methods play a key role in evaluating the performance of diagnostic tests, ensuring that they are accurate, reliable, and useful in clinical practice.

1.3.1 Sensitivity and Specificity

Sensitivity and specificity are two important statistical measures used to evaluate the performance of diagnostic tests. Sensitivity refers to a test's ability to correctly identify individuals who have the disease (true positives), while specificity measures the test's ability to correctly identify individuals who do not have the disease (true negatives). A highly sensitive test minimizes false negatives, whereas a highly specific test reduces false positives. These metrics help healthcare providers choose the most appropriate tests for different clinical scenarios.

1.3.2 Receiver Operating Characteristic (ROC) Curves

ROC curves are graphical tools used to assess the diagnostic performance of tests. By plotting the true positive rate (sensitivity) against the false positive rate (1-specificity), ROC curves allow researchers to visualize the trade-off between sensitivity and specificity. The area under the curve (AUC) is a summary measure of the test's overall accuracy. ROC analysis helps in selecting the optimal cutoff point for diagnostic tests and comparing the performance of different tests.

1.4 STATISTICAL METHODS IN GENOMICS AND PERSONALIZED MEDICINE

With the advent of high-throughput technologies in genomics, statistical methods have become indispensable in analyzing the vast amounts of data generated in this field. Genomic data, such as DNA sequences and gene expression levels, are used to identify genetic variations associated with diseases and to develop personalized treatment strategies.

1.4.1 Genome-Wide Association Studies (GWAS)

GWAS are studies that investigate the association between genetic variants and diseases by scanning the entire genome. Statistical techniques, such as logistic regression and chi-square tests, are used to analyze the relationships between single nucleotide polymorphisms (SNPs) and disease outcomes. GWAS have led to the discovery of genetic markers for various conditions, including cancer, diabetes, and heart disease.

1.4.2 Machine Learning in Genomics

Machine learning algorithms, such as decision trees, support vector machines, and neural networks, are increasingly being used in genomics to analyze complex datasets. These algorithms can identify patterns and relationships that may not be apparent through traditional statistical methods. Machine learning is particularly useful in personalized medicine, where it can predict individual responses to treatments based on genetic and clinical data.

1.5 STATISTICAL METHODS IN HEALTH POLICY AND PUBLIC HEALTH

Statistical methods are also applied in the field of public health and health policy to evaluate the effectiveness of health programs, allocate resources efficiently, and guide decision-making.

1.5.1 Cost-Effectiveness Analysis

Cost-effectiveness analysis (CEA) is a statistical technique used to assess the economic efficiency of healthcare interventions. By comparing the costs and health outcomes of different interventions, CEA helps policymakers decide how to allocate limited resources. For example, CEA can be used to determine whether a new drug provides sufficient health benefits relative to its cost compared to existing treatments.

1.5.2 Health Surveys and Data Collection

Public health agencies often conduct large-scale health surveys to collect data on the health status of populations. Statistical methods are used to design these surveys, ensuring that they are representative of the population and that the data collected are accurate and reliable. Techniques such as stratified sampling and weighting are employed to account for potential biases and ensure that the survey results can be generalized to the broader population.

1.6 CONCLUSION

Statistical methods are integral to every aspect of modern medicine, from clinical trials and epidemiology to genomics and public health policy. They provide the tools necessary for analyzing data, drawing valid conclusions, and making evidence-based decisions that improve patient care and public health outcomes. As medical research continues to evolve with new technologies and data sources, the scope of statistical methods in medicine will only continue to expand, offering new opportunities to advance the field and improve health worldwide.

CHAPTER 2

ROLE OF STATISTICS IN CLINICAL MEDICINE

Statistics play a vital role in clinical medicine, providing the foundation for evidence-based practice, research, and decision-making. As medicine becomes increasingly reliant on data to guide diagnosis, treatment, and policy decisions, statistical methods have emerged as indispensable tools in clinical research and patient care. From designing clinical trials to evaluating new treatments, interpreting diagnostic tests, and analyzing patient outcomes, statistics allow clinicians and researchers to draw meaningful conclusions from complex datasets.

2.1 ROLE OF STATISTICS IN CLINICAL TRIALS

Clinical trials are the gold standard for testing new medical treatments, drugs, and interventions. The design, execution, and analysis of clinical trials are heavily dependent on statistical methods. These methods ensure that the results are reliable, valid, and generalizable to larger populations, thereby providing the evidence needed for regulatory approval and clinical implementation.

2.1.1 Study Design and Randomization

The foundation of any clinical trial is its design. Statistical principles are used to develop the study's framework, ensuring that the trial can accurately assess the effect of an intervention. Randomization is a critical aspect of trial design, where participants are randomly assigned to different treatment groups to eliminate selection bias. This ensures that the groups are comparable in terms of demographic and clinical characteristics, and that the outcomes can be attributed to the treatment rather than other confounding factors. Statistical techniques such as stratified randomization or block randomization are often used to achieve balance between groups, especially in small or heterogeneous populations.

2.1.2 Sample Size Determination and Power Analysis

In clinical trials, determining the appropriate sample size is crucial to ensuring the study has enough statistical power to detect meaningful differences between treatment groups. Power analysis is a statistical method used to calculate the minimum sample size required to achieve a specified level of significance (commonly 0.05) and power (often 80% or 90%). If the sample size is too small, the study may fail to detect a true effect (Type II error), while an overly large sample size could waste resources. Proper sample size calculation helps researchers avoid these pitfalls and ensures that the trial is both ethical and efficient.

2.1.3 Hypothesis Testing and Statistical Inference

Clinical trials typically involve testing a hypothesis, such as whether a new treatment is superior to the standard care. Statistical tests, such as t-tests, chi-square tests, or analysis of variance (ANOVA), are used to compare outcomes between groups.

These tests allow researchers to make inferences about the population based on the sample data. Hypothesis testing helps determine whether observed differences are due to chance or reflect a true effect of the treatment. Statistical inference is further strengthened by calculating p-values, which quantify the probability that the observed result occurred by chance. A p-value below a predefined threshold (usually 0.05) indicates that the results are statistically significant.

2.1.4 Interpretation of Confidence Intervals

In addition to hypothesis testing, confidence intervals provide a range of values within which the true effect of a treatment is likely to lie. Confidence intervals give clinicians a sense of the precision and reliability of the trial results.

For example, a narrow confidence interval suggests a high level of precision, while a wide confidence interval indicates uncertainty about the true effect size. Confidence intervals are particularly important when assessing the clinical relevance of study findings, as they provide more information than p-values alone.

2.2 ROLE OF STATISTICS IN DIAGNOSTIC MEDICINE

Statistics are essential in evaluating the accuracy and reliability of diagnostic tests, which are central to clinical decision-making. Diagnostic tests must be rigorously assessed to ensure that they correctly identify patients with or without a disease. Statistical measures such as sensitivity, specificity, and predictive values help clinicians evaluate the performance of these tests.

2.2.1 Sensitivity and Specificity

Sensitivity refers to a test's ability to correctly identify individuals who have a disease (true positives), while specificity measures its ability to correctly identify individuals who do not have the disease (true negatives). Both sensitivity and specificity are crucial in determining the diagnostic accuracy of a test. For instance, a highly sensitive test is important for screening purposes, as it minimizes the number of false negatives, ensuring that most patients with the disease are identified. Conversely, a highly specific test is valuable for confirming a diagnosis, as it reduces the number of false positives.

2.2.2 Predictive Values and Likelihood Ratios

Predictive values are statistical measures used to assess the performance of a diagnostic test in a specific population. The positive predictive value (PPV) represents the probability that a patient with a positive test result actually has the disease, while the negative predictive value (NPV) reflects the probability that a patient with a negative result does not have the disease. These values depend on the prevalence of the disease in the population being tested. In addition, likelihood ratios are used to quantify how much a test result will change the odds of a patient having a disease. A likelihood ratio greater than 1 increases the probability of disease, while a ratio less than 1 decreases it.

2.2.3 Receiver Operating Characteristic (ROC) Curves

ROC curves are used to evaluate the performance of diagnostic tests by plotting the true positive rate (sensitivity) against the false positive rate (1-specificity) across different thresholds. The area under the ROC curve (AUC) is a measure of the test's overall accuracy, with values closer to 1 indicating higher accuracy. ROC analysis allows clinicians to choose the optimal cutoff point for a test based on the trade-off between sensitivity and specificity, thus improving diagnostic decision-making.

2.3 ROLE OF STATISTICS IN PROGNOSIS AND OUTCOME PREDICTION

Statistics play a crucial role in predicting patient outcomes and prognosis in clinical medicine. Predictive models, which incorporate multiple variables, help clinicians estimate the likelihood of specific outcomes, such as disease recurrence, survival, or treatment response.

2.3.1 Risk Prediction Models

Risk prediction models are statistical tools used to estimate an individual's risk of developing a disease or experiencing a particular outcome based on demographic, clinical, and genetic factors. For example, the Framingham Risk Score is a widely used model for estimating the risk of cardiovascular disease. These models are often developed using regression analysis, where multiple variables are assessed for their independent contribution to the outcome. By quantifying risk, these models help clinicians make informed decisions about preventive measures, treatment strategies, and patient counseling.

2.3.2 Survival Analysis

Survival analysis is a statistical method used to analyze time-to-event data, such as the time from diagnosis to death or disease recurrence. Techniques like the Kaplan-Meier estimator and Cox proportional hazards models are commonly employed to estimate survival probabilities and compare survival rates between different patient groups. These methods are especially useful in oncology, where they help assess the effectiveness of treatments and the progression of disease over time.

2.4 ROLE OF STATISTICS IN EVIDENCE-BASED MEDICINE

Evidence-based medicine (EBM) relies on the systematic review and analysis of clinical data to guide medical practice. Statistics are fundamental to the process of synthesizing evidence from multiple studies and determining the best course of action for patient care.

2.4.1 Meta-Analysis and Systematic Reviews

Meta-analysis is a statistical technique used to combine data from multiple studies to provide a more precise estimate of treatment effects. By pooling data from different clinical trials, meta-analysis increases the statistical power and allows for more robust conclusions. Systematic reviews use a rigorous, predefined methodology to assess the quality and findings of individual studies, often incorporating meta-analyses. These methods are key components of evidence-based medicine, as they help clinicians and policymakers make informed decisions based on the totality of available evidence.

2.4.2 Guidelines and Clinical Practice Recommendations

Statistical evidence derived from clinical trials and observational studies forms the basis of clinical practice guidelines and recommendations. These guidelines help standardize care and ensure that patients receive the most effective treatments. Statistical methods are used to grade the strength of evidence supporting each recommendation, guiding clinicians in their application of evidence to individual patient care.

2.5 ROLE OF STATISTICS IN PUBLIC HEALTH AND POLICY

Beyond individual patient care, statistics also play a critical role in public health and healthcare policy. Statistical analysis informs health interventions, resource allocation, and the evaluation of public health programs.

2.5.1 Health Surveillance and Data Analysis

Public health agencies rely on statistical methods to track disease incidence, monitor outbreaks, and evaluate the impact of interventions. Descriptive statistics and time-series analysis are used to monitor trends in disease prevalence and assess the effectiveness of vaccination programs, health campaigns, and other public health initiatives.

2.5.2 Cost-Effectiveness Analysis

In healthcare policy, cost-effectiveness analysis (CEA) is used to assess the value of medical interventions by comparing their costs and outcomes. CEA helps policymakers allocate resources efficiently, ensuring that healthcare dollars are spent on interventions that provide the greatest benefit for the population. Statistical methods are used to model health outcomes, estimate costs, and evaluate the overall impact of health interventions on public health.

2.6 CONCLUSION

Statistics are integral to every aspect of clinical medicine, from designing and analyzing clinical trials to assessing diagnostic tests, predicting patient outcomes, and guiding evidence-based practice. The use of rigorous statistical methods ensures that medical decisions are based on reliable data, leading to improved patient care and more effective public health interventions. As clinical medicine continues to evolve, the role of statistics will remain essential in advancing healthcare outcomes and improving the quality of medical research and practice.

CHAPTER 3

ROLE OF STATISTICS IN PREVENTIVE MEDICINE

Preventive medicine focuses on the prevention of diseases and health conditions before they occur, emphasizing the importance of health promotion and risk reduction. This proactive approach to healthcare relies heavily on statistical methods to assess the risks of disease, develop strategies for prevention, evaluate the effectiveness of interventions, and allocate resources. The role of statistics in preventive medicine is multifaceted, spanning areas such as epidemiology, public health surveillance, risk prediction, and program evaluation. By applying statistical tools to analyze health data, preventive medicine can identify trends, detect emerging health threats, and guide evidence-based decision-making.

3.1 STATISTICS IN EPIDEMIOLOGY

Epidemiology, the study of disease distribution and determinants in populations, forms the backbone of preventive medicine. Statistical methods are essential in epidemiology for analyzing patterns of health and disease, identifying risk factors, and understanding how various factors influence disease spread. Statistics help epidemiologists make informed decisions about intervention strategies, assess the burden of diseases, and track their evolution in different populations.

3.1.1 Descriptive Statistics in Disease Monitoring

Descriptive statistics play a critical role in summarizing health-related data, such as the incidence and prevalence of diseases. Incidence refers to the number of new cases of a disease within a specified time period, while prevalence refers to the total number of cases in a population at a given time. By using measures like these, public health professionals can understand the scope of health problems and allocate resources accordingly. For instance, high prevalence rates of a particular disease may signal the need for increased healthcare services and preventive programs in affected areas.

3.1.2 Identifying Risk Factors

Statistical methods such as regression analysis and multivariate modeling allow epidemiologists to identify and quantify the impact of various risk factors on the development of diseases. These methods help differentiate between variables that are merely associated with a disease and those that are causally linked. For example, through the use of logistic regression, epidemiologists can estimate the likelihood of individuals developing a condition based on factors such as age, lifestyle, genetic predisposition, or environmental exposures. This risk factor analysis is crucial for targeting preventive measures toward high-risk groups.

3.2 ROLE OF STATISTICS IN PUBLIC HEALTH SURVEILLANCE

Public health surveillance is a cornerstone of preventive medicine. It involves the continuous collection, analysis, and interpretation of health data to detect patterns and respond to emerging health issues. Statistical methods are vital for the real-time monitoring of public health, guiding interventions, and evaluating the impact of preventive measures.

3.2.1 Trend Analysis and Time-Series Data

Time-series analysis is a statistical technique used to monitor disease trends over time. It allows public health officials to observe seasonal patterns, fluctuations, and long-term trends in disease occurrence. For example, influenza surveillance relies on statistical models to predict seasonal peaks in flu cases, enabling healthcare systems to prepare for potential outbreaks. Similarly, time-series analysis helps track the effectiveness of vaccination campaigns, identify spikes in infectious diseases, and anticipate future healthcare needs.

3.2.2 Early Detection of Health Threats

Statistics also play a critical role in the early detection of emerging health threats, such as infectious disease outbreaks or environmental hazards. Statistical algorithms can be applied to public health data to detect unusual patterns or clusters of cases that might indicate the start of an epidemic.

These algorithms, often used in syndromic surveillance systems, analyze data from multiple sources, including emergency room visits, laboratory reports, and social media, to provide an early warning of potential outbreaks. Rapid detection allows for the timely implementation of preventive measures, such as quarantine or vaccination campaigns, to curb the spread of disease.

3.3 RISK PREDICTION AND HEALTH FORECASTING

One of the most valuable contributions of statistics to preventive medicine is its role in risk prediction and health forecasting. Statistical models are used to estimate the probability of individuals developing specific diseases or health conditions based on known risk factors.

These predictive models help clinicians and public health professionals tailor preventive strategies to individual patients or populations, thereby optimizing healthcare interventions.

3.3.1 Risk Prediction Models

Risk prediction models are statistical tools that estimate an individual's likelihood of developing a disease or experiencing a health event based on demographic, clinical, and genetic factors.

For example, the Framingham Risk Score is a widely used tool for predicting the risk of cardiovascular disease. By incorporating variables such as age, cholesterol levels, blood pressure, and smoking status, the model calculates a patient's risk of having a heart attack or stroke over a given period. These models are instrumental in preventive medicine, as they allow healthcare providers to offer targeted advice on lifestyle modifications, medications, or other interventions to reduce the risk of disease.

3.3.2 Health Forecasting

Health forecasting involves using statistical models to predict future health trends based on historical data and current patterns. This is particularly important for planning preventive measures and healthcare resource allocation. For example, forecasting models are used to predict the burden of chronic diseases such as diabetes, hypertension, and cancer, allowing policymakers to implement public health campaigns, screening programs, and preventive treatments in anticipation of rising disease incidence. In infectious disease control, health forecasting helps guide vaccination policies, antibiotic use, and infection prevention strategies based on expected disease outbreaks.

3.4 EVALUATING PREVENTIVE INTERVENTIONS

Evaluating the effectiveness of preventive interventions is a key aspect of preventive medicine. Statistical methods are used to assess whether interventions such as vaccination programs, health education campaigns, and screening tests achieve their intended outcomes. This evaluation process helps refine preventive strategies and ensure that they are both effective and cost-efficient.

3.4.1 Clinical Trials and Preventive Interventions

Randomized controlled trials (RCTs) are the gold standard for evaluating the efficacy of preventive interventions. Statistical methods are employed throughout the design and analysis of these trials to ensure that the findings are robust and generalizable. For example, an RCT might be conducted to assess the effectiveness of a new vaccine in preventing a disease. Statistical techniques such as survival analysis and hypothesis testing are used to compare outcomes between the vaccinated group and a control group. The results of such trials inform public health recommendations and vaccination policies, ensuring that preventive interventions are evidence-based.

3.4.2 Cost-Effectiveness Analysis

Cost-effectiveness analysis (CEA) is a statistical tool used to evaluate the economic efficiency of preventive interventions. In healthcare, resources are often limited, and CEA helps policymakers prioritize interventions that provide the greatest benefit for the least cost. By comparing the costs and health outcomes of different preventive strategies, CEA ensures that healthcare budgets are spent efficiently. For example, a CEA might compare the cost per life saved between two cancer screening programs, guiding decision-makers on which program to implement more widely. This type of analysis is crucial in preventive medicine, where the goal is to maximize health outcomes at the population level while minimizing costs.

3.5 STATISTICS IN VACCINATION AND IMMUNIZATION PROGRAMS

Vaccination is one of the most successful public health interventions in preventive medicine. Statistical methods play a pivotal role in designing, implementing, and evaluating vaccination and immunization programs. From determining vaccine efficacy to monitoring adverse events and measuring herd immunity, statistics are essential in ensuring the success of these programs.

3.5.1 Vaccine Efficacy and Effectiveness

Statistical analysis is used to measure vaccine efficacy in clinical trials and vaccine effectiveness in real-world settings. Vaccine efficacy refers to the reduction in disease incidence among vaccinated individuals compared to unvaccinated ones in controlled trials, while vaccine effectiveness measures the real-world impact of vaccination programs. Statistical methods such as logistic regression and risk ratio calculation are used to estimate these measures, providing the evidence needed to recommend vaccines for widespread use.

3.5.2 Herd Immunity and Coverage Rates

Herd immunity occurs when a significant portion of a population is immune to a disease, either through vaccination or natural infection, reducing the likelihood of disease spread. Statistical models are used to calculate the vaccination coverage rates required to achieve herd immunity for various diseases. For example, the threshold for herd immunity against measles is estimated to be around 95%, meaning that 95% of the population needs to be vaccinated to prevent outbreaks. These models guide immunization strategies and help policymakers set targets for vaccination campaigns.

3.6 STATISTICS IN SCREENING AND EARLY DETECTION

Early detection of diseases through screening is a key component of preventive medicine. Screening programs aim to identify diseases at an early, more treatable stage, reducing morbidity and mortality. Statistical methods are used to assess the accuracy, reliability, and effectiveness of screening tests.

3.6.1 Sensitivity, Specificity, and Predictive Values

The performance of a screening test is measured using statistical metrics such as sensitivity, specificity, positive predictive value (PPV), and negative predictive value (NPV). Sensitivity measures a test's ability to correctly identify individuals with a disease (true positives), while specificity measures its ability to correctly identify individuals without the disease (true negatives). PPV and NPV indicate the probability that individuals with positive or negative test results truly have or do not have the disease. These metrics are crucial for determining whether a screening test is suitable for widespread use in a population.

3.6.2 Receiver Operating Characteristic (ROC) Curves

ROC curves are statistical tools used to assess the diagnostic accuracy of screening tests by plotting the true positive rate (sensitivity) against the false positive rate (1-specificity) at various threshold levels. The area under the curve (AUC) provides an overall measure of test performance, with values closer to 1 indicating higher accuracy. ROC analysis helps optimize screening strategies by determining the best cutoff points for identifying diseases.

3.7 CONCLUSION

Statistics are an integral part of preventive medicine, supporting every aspect of disease prevention, health promotion, and risk reduction. From epidemiology and public health surveillance to risk prediction, intervention evaluation, and vaccination programs, statistical methods provide the tools needed to make informed decisions and develop effective preventive strategies. By applying rigorous statistical analysis to health data, preventive medicine can improve population health outcomes, reduce disease burden, and optimize healthcare resources. As healthcare continues to evolve, the role of statistics in preventive medicine will remain essential in addressing current and future public health challenges.

CHAPTER 4
VARIABLES

In biostatistics and research methods, variables play a fundamental role in the design, analysis, and interpretation of data. A variable is any measurable trait or characteristic that can take on different values, such as height, age, blood pressure, or cholesterol levels. Variables allow researchers to quantify and analyze phenomena, determine relationships between factors, and make inferences about populations based on sample data.

4.1 TYPES OF VARIABLES

4.1.1 Independent and Dependent Variable

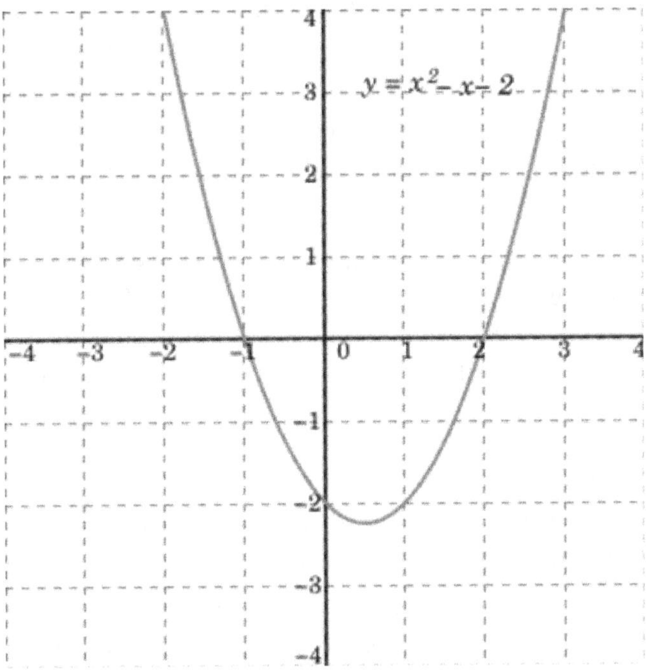

Fig 1: In Single Variable Calculus, a Function is Typically Graphed with the Horizontal Axis Representing the Independent Variable and the Vertical Axis Representing the Dependent Variable. in this Function, *y* is the Dependent Variable and *x* is the Independent Variable

A variable is considered dependent if it depends on an independent variable. Dependent variables are studied under the supposition or demand that they depend, by some law or rule (e.g., by a mathematical function), on the values of other variables. Independent variables, in turn, are not seen as depending on any other variable in the scope of the experiment in question. In this sense, some common independent variables are time, space, density, mass, fluid flow rate, and previous values of some observed value of interest (e.g. human population size) to predict future values (the dependent variable).

Of the two, it is always the dependent variable whose variation is being studied, by altering inputs, also known as regressors in a statistical context. In an experiment, any variable that can be attributed a value without attributing a value to any other variable is called an independent variable.

Models and experiments test the effects that the independent variables have on the dependent variables. Sometimes, even if their influence is not of direct interest, independent variables may be included for other reasons, such as to account for their potential confounding effect.

In Pure Mathematics

In mathematics, a function is a rule for taking an input (in the simplest case, a number or set of numbers) and providing an output (which may also be a number). A symbol that stands for an arbitrary input is called an independent variable, while a symbol that stands for an arbitrary output is called a dependent variable. The most common symbol for the input is x, and the most common symbol for the output is y; the function itself is commonly written $y = f(x)$.

It is possible to have multiple independent variables or multiple dependent variables. For instance, in multivariable calculus, one often encounters functions of the form $z = f(x,y)$, where z is a dependent variable and x and y are independent variables. Functions with multiple outputs are often referred to as vector-valued functions.

In Modeling and Statistics

In mathematical modeling, the relationship between the set of dependent variables and set of independent variables is studied.

In the simple stochastic linear model $y_i = a + bx_i + e_i$ the term y_i is the ith value of the dependent variable and x_i is the ith value of the independent variable. The term e_i is known as the "error" and contains the variability of the dependent variable not explained by the independent variable.

With multiple independent variables, the model is $y_i = a + bx_{i,1} + bx_{i,2} + ... + bx_{i,n} + e_i$, where n is the number of independent variables.

In statistics, more specifically in linear regression, a scatter plot of data is generated with X as the independent variable and Y as the dependent variable. This is also called a bivariate dataset, $(x_1, y_1)(x_2, y_2) \ldots (x_i, y_i)$. The simple linear regression model takes the form of $Y_i = a + Bx_i + U_i$, for $i = 1, 2, \ldots, n$. In this case, $U_i, \ldots, U_n$ are independent random variables. This occurs when the measurements do not influence each other. Through propagation of independence, the independence of U_i implies independence of Y_i, even though each Y_i has a different expectation value. Each U_i has an expectation value of 0 and a variance of σ^2. Expectation of Y_i Proof:

$$E[Y_i] = E[\alpha + \beta x_i + U_i] = \alpha + \beta x_i + E[U_i] = \alpha + \beta x_i.$$

The line of best fit for the bivariate dataset takes the form $y = \alpha + \beta x$ and is called the regression line. α and β correspond to the intercept and slope, respectively.

In an experiment, the variable manipulated by an experimenter is something that is proven to work, called an independent variable. The dependent variable is the event expected to change when the independent variable is manipulated.

In data mining tools (for multivariate statistics and machine learning), the dependent variable is assigned a *role* as target variable (or in some tools as *label attribute*), while an independent variable may be assigned a role as *regular variable.* Known values for the target variable are provided for the training data set and test data set, but should be predicted for other data. The target variable is used in supervised learning algorithms but not in unsupervised learning.

Synonyms

Depending on the context, an independent variable is sometimes called a "predictor variable", "regressor", "covariate", "manipulated variable", "explanatory variable", "exposure variable", "risk factor", "feature" (in machine learning and pattern recognition) or "input variable". In econometrics, the term "control variable" is usually used instead of "covariate".

"Explanatory variable" is preferred by some authors over "independent variable" when the quantities treated as independent variables may not be statistically independent or independently manipulable by the researcher. If the independent variable is referred to as an "explanatory variable" then the term "response variable" is preferred by some authors for the dependent variable.

Depending on the context, a dependent variable is sometimes called a "response variable", "regressand", "criterion", "predicted variable", "measured variable", "explained variable", "experimental variable", "responding variable", "outcome variable", "output variable", "target" or "label". In economics endogenous variables are usually referencing the target.

"Explained variable" is preferred by some authors over "dependent variable" when the quantities treated as "dependent variables" may not be statistically dependent.

If the dependent variable is referred to as an "explained variable" then the term "predictor variable" is preferred by some authors for the independent variable.

An example is provided by the analysis of trend in sea level by Woodworth (1987). Here the dependent variable (and variable of most interest) was the annual mean sea level at a given location for which a series of yearly values were available. The primary independent variable was time. Use was made of a covariate consisting of yearly values of annual mean atmospheric pressure at sea level.

Other Variables

A variable may be thought to alter the dependent or independent variables, but may not actually be the focus of the experiment. So that the variable will be kept constant or monitored to try to minimize its effect on the experiment. Such variables may be designated as either a "controlled variable", "control variable", or "fixed variable".

Extraneous variables, if included in a regression analysis as independent variables, may aid a researcher with accurate response parameter estimation, prediction, and goodness of fit, but are not of substantive interest to the hypothesis under examination. For example, in a study examining the effect of post-secondary education on lifetime earnings, some extraneous variables might be gender, ethnicity, social class, genetics, intelligence, age, and so forth. A variable is extraneous only when it can be assumed (or shown) to influence the dependent variable. If included in a regression, it can improve the fit of the model. If it is excluded from the regression and if it has a non-zero covariance with one or more of the independent variables of interest, its omission will bias the regression's result for the effect of that independent variable of interest. This effect is called confounding or omitted variable bias; in these situations, design changes and/or controlling for a variable statistical control is necessary.

Extraneous variables are often classified into three types:

- Subject variables, which are the characteristics of the individuals being studied that might affect their actions. These variables include age, gender, health status, mood, background, etc.
- Blocking variables or experimental variables are characteristics of the persons conducting the experiment which might influence how a person behaves. Gender, the presence of racial discrimination, language, or other factors may qualify as such variables.
- Situational variables are features of the environment in which the study or research was conducted, which have a bearing on the outcome of the experiment in a negative way. Included are the air temperature, level of activity, lighting, and time of day.

In modelling, variability that is not covered by the independent variable is designated by e I e_I and is known as the "residual", "side effect", "error", "unexplained share", "residual variable", "disturbance", or "tolerance".

Examples

- **Effect of Fertilizer on Plant Growths:** In a study measuring the influence of different quantities of fertilizer on plant growth, the independent variable would be the amount of fertilizer used. The dependent variable would be the growth in height or mass of the plant. The controlled variables would be the type of plant, the type of fertilizer, the amount of sunlight the plant gets, the size of the pots, etc.
- **Effect of Drug Dosage on Symptom Severity:** In a study of how different doses of a drug affect the severity of symptoms, a researcher could compare the frequency and intensity of symptoms when different doses are administered. Here the independent variable is the dose and the dependent variable is the frequency/intensity of symptoms.
- **Effect of Temperature on Pigmentation:** In measuring the amount of color removed from beetroot samples at different temperatures, temperature is the independent variable and amount of pigment removed is the dependent variable.
- **Effect of Sugar Added in a Coffee:** The taste varies with the amount of sugar added in the coffee. Here, the sugar is the independent variable, while the taste is the dependent variable.

4.1.2 Continuous Variables

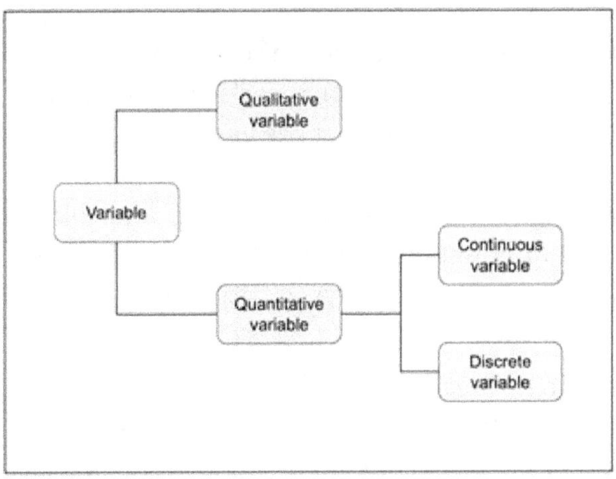

Fig 2: Variables can be Divided into Two Main Categories: Qualitative (Categorical) and Quantitative (Numerical).

In mathematics and statistics, a quantitative variable may be continuous or discrete if they are typically obtained by *measuring* or *counting*, respectively. If it can take on two particular real values such that it can also take on all real values between them (including values that are arbitrarily or infinitesimally close together), the variable is continuous in that interval. If it can take on a value such that there is a non-infinitesimal gap on each side of it containing no values that the variable can take on, then it is discrete around that value. In some contexts, a variable can be discrete in some ranges of the number line and continuous in others.

Continuous Variable

A continuous variable is a variable such that there are possible values between any two values.

For example, a variable over a non-empty range of the real numbers is continuous, if it can take on any value in that range.

Methods of calculus are often used in problems in which the variables are continuous, for example in continuous optimization problems.

In statistical theory, the probability distributions of continuous variables can be expressed in terms of probability density functions.

In continuous-time dynamics, the variable *time* is treated as continuous, and the equation describing the evolution of some variable over time is a differential equation. The instantaneous rate of change is a well-defined concept that takes the ratio of the change in the dependent variable to the independent variable at a specific instant.

Discrete Variable

In contrast, a variable is a discrete variable if and only if there exists a one-to-one correspondence between this variable and a subset of N the set of natural numbers. In other words, a discrete variable over a particular interval of real values is one for which, for any value in the range that the variable is permitted to take on, there is a positive minimum distance to the nearest other permissible value. The value of a discrete variable can be obtained by counting, and the number of permitted values is either finite or countably infinite. Common examples are variables that must be integers, non-negative integers, positive integers, or only the integers 0 and 1.

Methods of calculus do not readily lend themselves to problems involving discrete variables. Especially in multivariable calculus, many models rely on the assumption of continuity. Examples of problems involving discrete variables include integer programming. In statistics, the probability distributions of discrete variables can be expressed in terms of probability mass functions. In discrete time dynamics, the variable *time* is treated as discrete, and the equation of evolution of some variable over time is called a difference equation. For certain discrete-time dynamical systems, the system response can be modelled by solving the difference equation for an analytical solution.

In econometrics and more generally in regression analysis, sometimes some of the variables being empirically related to each other are 0-1 variables, being permitted to take on only those two values. The purpose of the discrete values of 0 and 1 is to use the dummy variable as a 'switch' that can 'turn on' and 'turn off' by assigning the two values to different parameters in an equation. A variable of this type is called a dummy variable. If the dependent variable is a dummy variable, then logistic regression or probit regression is commonly employed. In the case of regression analysis, a dummy variable can be used to represent subgroups of the sample in a study (e.g. the value 0 corresponding to a constituent of the control group).

Mixture of Continuous and Discrete Variables

A mixed multivariate model can contain both discrete and continuous variables. For instance, a simple mixed multivariate model could have a discrete variable x x, which only takes on values 0 or 1, and a continuous variable y y. An example of a mixed model could be a research study on the risk of psychological disorders based on one binary measure of psychiatric symptoms and one continuous measure of cognitive performance. Mixed models may also involve a single variable that is discrete over some range of the number line and continuous at another range.

In probability theory and statistics, the probability distribution of a mixed random variable consists of both discrete and continuous components. A mixed random variable does not have a cumulative distribution function that is discrete or everywhere-continuous. An example of a mixed type random variable is the probability of wait time in a queue. The likelihood of a customer experiencing a zero wait time is discrete, while non-zero wait times are evaluated on a continuous time scale. In physics (particularly quantum mechanics, where this sort of distribution often arises), dirac delta functions are often used to treat continuous and discrete components in a unified manner. For example, the previous example might be described by a probability density

$$p(t) = \alpha\delta(t) + g(t), such that\ P(t > 0) = \int_0^\infty g(t) = 1 - \alpha,\ and\ P(t = 0) = \alpha$$

4.1.3 Categorical Variables

In statistics, a categorical variable (also called qualitative variable) is a variable that can take on one of a limited, and usually fixed, number of possible values, assigning each individual or other unit of observation to a particular group or nominal category on the basis of some qualitative property. In computer science and some branches of mathematics, categorical variables are referred to as enumerations or enumerated types. Commonly each of the possible values of a categorical variable is referred to as a level. The probability distribution associated with a random categorical variable is called a categorical distribution.

Categorical data is the statistical data type consisting of categorical variables or of data that has been converted into that form, for example as grouped data. More specifically, categorical data may derive from observations made of qualitative data that are summarised as counts or cross tabulations, or from observations of quantitative data grouped within given intervals. Often, purely categorical data are summarised in the form of a contingency table. However, particularly when considering data analysis, it is common to use the term "categorical data" to apply to data sets that, while containing some categorical variables, may also contain non-categorical variables. Ordinal variables have a meaningful ordering, while nominal variables have no meaningful ordering.

A categorical variable that can take on exactly two values is termed a *binary variable* or a dichotomous variable; an important special case is the Bernoulli variable. Categorical variables with more than two possible values are called polytomous variables; categorical variables are often assumed to be polytomous unless otherwise specified. Discretization is treating continuous data as if it were categorical. Dichotomization is treating continuous data or polytomous variables as if they were binary variables. Regression analysis often treats category membership with one or more quantitative dummy variables.

Examples of Categorical Variables

Examples of values that might be represented in a categorical variable:

- The roll of a six-sided dice: possible outcomes are 1, 2, 3, 4, 5, or 6.
- Demographic information of a population: gender, disease status.
- The blood type of a person: A, B, AB or O.
- The political party that a voter might vote for, e. g. *Green Party*, *Christian Democrat*, *Social Democrat*, etc.
- The type of a rock: igneous, sedimentary or metamorphic.
- The identity of a particular word (e.g., in a language model): One of V possible choices, for a vocabulary of size V.

Notation

For ease in statistical processing, categorical variables may be assigned numeric indices, e.g. 1 through K for a K-way categorical variable (i.e. a variable that can express exactly K possible values). In general, however, the numbers are arbitrary, and have no significance beyond simply providing a convenient label for a particular value. In other words, the values in a categorical variable exist on a nominal scale: they each represent a logically separate concept, cannot necessarily be meaningfully ordered, and cannot be otherwise manipulated as numbers could be. Instead, valid operations are equivalence, set membership, and other set-related operations.

As a result, the central tendency of a set of categorical variables is given by its mode; neither the mean nor the median can be defined. As an example, given a set of people, we can consider the set of categorical variables corresponding to their last names. We can consider operations such as equivalence (whether two people have the same last name), set membership (whether a person has a name in a given list), counting (how many people have a given last name), or finding the mode (which name occurs most often). However, we cannot meaningfully compute the "sum" of Smith + Johnson, or ask whether Smith is "less than" or "greater than" Johnson. As a result, we cannot meaningfully ask what the "average name" (the mean) or the "middle-most name" (the median) is in a set of names.

This ignores the concept of alphabetical order, which is a property that is not inherent in the names themselves, but in the way we construct the labels. For example, if we write the names in Cyrillic and consider the Cyrillic ordering of letters, we might get a different result of evaluating "Smith < Johnson" than if we write the names in the standard Latin alphabet; and if we write the names in Chinese characters, we cannot meaningfully evaluate "Smith < Johnson" at all, because no consistent ordering is defined for such characters. However, if we do consider the names as written, e.g., in the Latin alphabet, and define an ordering corresponding to standard alphabetical order, then we have effectively converted them into ordinal variables defined on an ordinal scale.

Number of Possible Values

Categorical random variables are normally described statistically by a categorical distribution, which allows an arbitrary *K*-way categorical variable to be expressed with separate probabilities specified for each of the *K* possible outcomes. Such multiple-category categorical variables are often analyzed using a multinomial distribution, which counts the frequency of each possible combination of numbers of occurrences of the various categories. Regression analysis on categorical outcomes is accomplished through multinomial logistic regression, multinomial probit or a related type of discrete choice model.

Categorical variables that have only two possible outcomes (e.g., "yes" vs. "no" or "success" vs. "failure") are known as *binary variables* (or *Bernoulli variables*). Because of their importance, these variables are often considered a separate category, with a separate distribution (the Bernoulli distribution) and separate regression models (logistic regression, probit regression, etc.). As a result, the term "categorical variable" is often reserved for cases with 3 or more outcomes, sometimes termed a *multi-way* variable in opposition to a binary variable.

It is also possible to consider categorical variables where the number of categories is not fixed in advance. As an example, for a categorical variable describing a particular word, we might not know in advance the size of the vocabulary, and we would like to allow for the possibility of encountering words that we have not already seen. Standard statistical models, such as those involving the categorical distribution and multinomial logistic regression, assume that the number of categories is known in advance, and changing the number of categories on the fly is tricky.

In such cases, more advanced techniques must be used. An example is the Dirichlet process, which falls in the realm of nonparametric statistics. In such a case, it is logically assumed that an infinite number of categories exist, but at any one time most of them (in fact, all but a finite number) have never been seen. All formulas are phrased in terms of the number of categories actually seen so far rather than the (infinite) total number of potential categories in existence, and methods are created for incremental updating of statistical distributions, including adding "new" categories.

Categorical Variables and Regression

Categorical variables represent a qualitative method of scoring data (i.e. represents categories or group membership). These can be included as independent variables in a regression analysis or as dependent variables in logistic regression or probit regression, but must be converted to quantitative data in order to be able to analyze the data. One does so through the use of coding systems. Analyses are conducted such that only g -1 (g being the number of groups) are coded. This minimizes redundancy while still representing the complete data set as no additional information would be gained from coding the total g groups: for example, when coding gender (where $g = 2$: male and female), if we only code females everyone left over would necessarily be males. In general, the group that one does not code for is the group of least interest.

There are three main coding systems typically used in the analysis of categorical variables in regression: dummy coding, effects coding, and contrast coding. The regression equation takes the form of $Y = bX + a$, where b is the slope and gives the weight empirically assigned to an explanator, X is the explanatory variable, and a is the Y-intercept, and these values take on different meanings based on the coding system used. The choice of coding system does not affect the F or R^2 statistics. However, one chooses a coding system based on the comparison of interest since the interpretation of b values will vary.

Dummy Coding

Dummy coding is used when there is a control or comparison group in mind. One is therefore analyzing the data of one group in relation to the comparison group: *a* represents the mean of the control group and *b* is the difference between the mean of the experimental group and the mean of the control group. It is suggested that three criteria be met for specifying a suitable control group: the group should be a well-established group (e.g. should not be an "other" category), there should be a logical reason for selecting this group as a comparison (e.g. the group is anticipated to score highest on the dependent variable), and finally, the group's sample size should be substantive and not small compared to the other groups.

In dummy coding, the reference group is assigned a value of 0 for each code variable, the group of interest for comparison to the reference group is assigned a value of 1 for its specified code variable, while all other groups are assigned 0 for that particular code variable.

The *b* values should be interpreted such that the experimental group is being compared against the control group. Therefore, yielding a negative b value would entail the experimental group have scored less than the control group on the dependent variable. To illustrate this, suppose that we are measuring optimism among several nationalities and we have decided that French people would serve as a useful control. If we are comparing them against Italians, and we observe a negative *b* value, this would suggest Italians obtain lower optimism scores on average.

The following table is an example of dummy coding with *French* as the control group and C1, C2, and C3 respectively being the codes for *Italian*, *German*, and *Other* (neither French nor Italian nor German):

Nationality	C1	C2	C3
French	0	0	0
Italian	1	0	0
German	0	1	0
Other	0	0	1

Effects Coding

In the effects coding system, data are analyzed through comparing one group to all other groups. Unlike dummy coding, there is no control group. Rather, the comparison is being made at the mean of all groups combined (*a* is now the grand mean). Therefore, one is not looking for data in relation to another group but rather, one is seeking data in relation to the grand mean. Effects coding can either be weighted or unweighted. Weighted effects coding is simply calculating a weighted grand mean, thus taking into account the sample size in each variable. This is most appropriate in situations where the sample is representative of the population in question. Unweighted effects coding is most appropriate in situations where differences in sample size are the result of incidental factors. The interpretation of *b* is different for each: in unweighted effects coding *b* is the difference between the mean of the experimental group and the grand mean, whereas in the weighted situation it is the mean of the experimental group minus the weighted grand mean.

In effects coding, we code the group of interest with a 1, just as we would for dummy coding. The principal difference is that we code -1 for the group we are least interested in. Since we continue to use a *g* - 1 coding scheme, it is in fact the -1 coded group that will not produce data, hence the fact that we are least interested in that group. A code of 0 is assigned to all other groups.

The *b* values should be interpreted such that the experimental group is being compared against the mean of all groups combined (or weighted grand mean in the case of weighted effects coding). Therefore, yielding a negative *b* value would entail the coded group as having scored less than the mean of all groups on the dependent variable. Using our previous example of optimism scores among nationalities, if the group of interest is Italians, observing a negative *b* value suggest they obtain a lower optimism score. The following table is an example of effects coding with *Other* as the group of least interest.

Nationality	C1	C2	C3
French	0	0	1
Italian	1	0	0
German	0	1	0
Other	−1	−1	−1

Contrast Coding

The contrast coding system allows a researcher to directly ask specific questions. Rather than having the coding system dictate the comparison being made (i.e., against a control group as in dummy coding, or against all groups as in effects coding) one can design a unique comparison catering to one's specific research question.

This tailored hypothesis is generally based on previous theory and/or research. The hypotheses proposed are generally as follows: first, there is the central hypothesis which postulates a large difference between two sets of groups; the second hypothesis suggests that within each set, the differences among the groups are small. Through its a priori focused hypotheses, contrast coding may yield an increase in power of the statistical test when compared with the less directed previous coding systems.

Certain differences emerge when we compare our a priori coefficients between ANOVA and regression. Unlike when used in ANOVA, where it is at the researcher's discretion whether they choose coefficient values that are either orthogonal or non-orthogonal, in regression, it is essential that the coefficient values assigned in contrast coding be orthogonal. Furthermore, in regression, coefficient values must be either in fractional or decimal form. They cannot take on interval values.

The construction of contrast codes is restricted by three rules:

- The sum of the contrast coefficients per each code variable must equal zero.
- The difference between the sum of the positive coefficients and the sum of the negative coefficients should equal 1.
- Coded variables should be orthogonal.

Violating rule 2 produces accurate R^2 and F values, indicating that we would reach the same conclusions about whether or not there is a significant difference; however, we can no longer interpret the *b* values as a mean difference.

To illustrate the construction of contrast codes consider the following table. Coefficients were chosen to illustrate our a priori hypotheses: Hypothesis 1: French and Italian persons will score higher on optimism than Germans (French = +0.33, Italian = +0.33, German = -0.66).

This is illustrated through assigning the same coefficient to the French and Italian categories and a different one to the Germans.

The signs assigned indicate the direction of the relationship (hence giving Germans a negative sign is indicative of their lower hypothesized optimism scores). Hypothesis 2: French and Italians are expected to differ on their optimism scores (French = +0.50, Italian = -0.50, German = 0). Here, assigning a zero value to Germans demonstrates their non-inclusion in the analysis of this hypothesis. Again, the signs assigned are indicative of the proposed relationship.

Nationality	C1	C2
French	+0.33	+0.50
Italian	+0.33	−0.50
German	−0.66	0

Nonsense Coding

Nonsense coding occurs when one uses arbitrary values in place of the designated "0"s "1"s and "-1"s seen in the previous coding systems. Although it produces correct mean values for the variables, the use of nonsense coding is not recommended as it will lead to uninterpretable statistical results.

Embeddings

Embeddings are codings of categorical values into low-dimensional real-valued (sometimes complex-valued) vector spaces, usually in such a way that 'similar' values are assigned 'similar' vectors, or with respect to some other kind of criterion making the vectors useful for the respective application. A common special case are word embeddings, where the possible values of the categorical variable are the words in a language and words with similar meanings are to be assigned similar vectors.

Interactions

An interaction may arise when considering the relationship among three or more variables, and describes a situation in which the simultaneous influence of two variables on a third is not additive. Interactions may arise with categorical variables in two ways: either categorical by categorical variable interactions, or categorical by continuous variable interactions.

Categorical by Categorical Variable Interactions

This type of interaction arises when we have two categorical variables. In order to probe this type of interaction, one would code using the system that addresses the researcher's hypothesis most appropriately. The product of the codes yields the interaction. One may then calculate the *b* value and determine whether the interaction is significant.

Categorical by Continuous Variable Interactions

Simple slopes analysis is a common post hoc test used in regression which is similar to the simple effects analysis in ANOVA, used to analyze interactions. In this test, we are examining the simple slopes of one independent variable at specific values of the other independent variable. Such a test is not limited to use with continuous variables, but may also be employed when the independent variable is categorical. We cannot simply choose values to probe the interaction as we would in the continuous variable case because of the nominal nature of the data (i.e., in the continuous case, one could analyze the data at high, moderate, and low levels assigning 1 standard deviation above the mean, at the mean, and at one standard deviation below the mean respectively). In our categorical case we would use a simple regression equation for each group to investigate the simple slopes. It is common practice to standardize or center variables to make the data more interpretable in simple slopes analysis; however, categorical variables should never be standardized or centered. This test can be used with all coding systems.

4.1.4 Confounding Variables

In causal inference, a confounder is a variable that influences both the dependent variable and independent variable, causing a spurious association. Confounding is a causal concept, and as such, cannot be described in terms of correlations or associations.

The existence of confounders is an important quantitative explanation why correlation does not imply causation. Some notations are explicitly designed to identify the existence, possible existence, or non-existence of confounders in causal relationships between elements of a system.

Simple Example

Let's assume that a trucking company owns a fleet of trucks made by two different manufacturers. Trucks made by one manufacturer are called "A Trucks" and trucks made by the other manufacturer are called "B Trucks." We want to find out whether A Trucks or B Trucks get better fuel economy.

We measure fuel and miles driven for a month and calculate the MPG for each truck. We then run the appropriate analysis, which determines that there is a statistically significant trend that A Trucks are more fuel efficient than B Trucks. Upon further reflection, however, we also notice that A Trucks are more likely to be assigned highway routes, and B Trucks are more likely to be assigned city routes. This is a confounding variable.

The confounding variable makes the results of the analysis unreliable. It is quite likely that we are just measuring the fact that highway driving results in better fuel economy than city driving.

In statistics terms, the make of the truck is the independent variable, the fuel economy (MPG) is the dependent variable and the amount of city driving is the confounding variable.

To fix this study, we have several choices. One is to randomize the truck assignments so that A trucks and B Trucks end up with equal amounts of city and highway driving.

That eliminates the confounding variable. Another choice is to quantify the amount of city driving and use that as a second independent variable. A third choice is to segment the study, first comparing MPG during city driving for all trucks, and then run a separate study comparing MPG during highway driving.

Control

Consider a researcher attempting to assess the effectiveness of drug *X*, from population data in which drug usage was a patient's choice. The data shows that gender (*Z*) influences a patient's choice of drug as well as their chances of recovery (*Y*). In this scenario, gender *Z* confounds the relation between X and Y since *Z* is a cause of both *X* and *Y*:

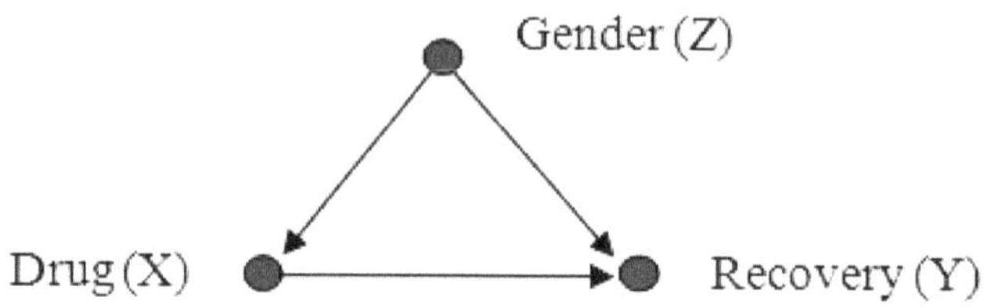

We have that

$$P(y \mid \text{do}(x)) \neq P(y \mid x)$$

because the observational quantity contains information about the correlation between X and Z, and the interventional quantity does not (since X is not correlated with Z in a randomized experiment). It can be shown that, in cases where only observational data is available, an unbiased estimate of the desired quantity $P(y \mid \text{do}(x))$, can be obtained by "adjusting" for all confounding factors, namely, conditioning on their various values and averaging the result.

In the case of a single confounder Z, this leads to the "adjustment formula":

$$P(y \mid \text{do}(x)) = \sum_z P(y \mid x, z) P(z)$$

which gives an unbiased estimate for the causal effect of X on Y. The same adjustment formula works when there are multiple confounders except, in this case, the choice of a set Z of variables that would guarantee unbiased estimates must be done with caution. The criterion for a proper choice of variables is called the Back-Door and requires that the chosen set Z "blocks" (or intercepts) every path between X and Y that contains an arrow into X. Such sets are called "Back-Door admissible" and may include variables which are not common causes of X and Y, but merely proxies thereof.

History

According to Morabia (2011), the word *confounding* derives from the Medieval Latin verb "confundere", which meant "mixing", and was probably chosen to represent the confusion (from Latin: con=with + fusus=mix or fuse together) between the cause one wishes to assess and other causes that may affect the outcome and thus confuse, or stand in the way of the desired assessment. Greenland, Robins and Pearl note an early use of the term "confounding" in causal inference by John Stuart Mill in 1843.

Fisher introduced the word "confounding" in his 1935 book "The Design of Experiments" to refer specifically to a consequence of blocking (i.e., partitioning) the set of treatment combinations in a factorial experiment, whereby certain interactions may be "confounded with blocks". This popularized the notion of confounding in statistics, although Fisher was concerned with the control of heterogeneity in experimental units, not with causal inference.

According to Vandenbroucke (2004) it was Kish who used the word "confounding" in the sense of "incomparability" of two or more groups (e.g., exposed and unexposed) in an observational study. Formal conditions defining what makes certain groups "comparable" and others "incomparable" were later developed in epidemiology by Greenland and Robins (1986) using the counterfactual language of Neyman (1935) and Rubin (1974). These were later supplemented by graphical criteria such as the Back-Door condition (Pearl 1993; Greenland, Robins and Pearl 1999).

Graphical criteria were shown to be formally equivalent to the counterfactual definition but more transparent to researchers relying on process models.

Types

In the case of risk assessments evaluating the magnitude and nature of risk to human health, it is important to control for confounding to isolate the effect of a particular hazard such as a food additive, pesticide, or new drug. For prospective studies, it is difficult to recruit and screen for volunteers with the same background (age, diet, education, geography, etc.), and in historical studies, there can be similar variability. Due to the inability to control for variability of volunteers and human studies, confounding is a particular challenge. For these reasons, experiments offer a way to avoid most forms of confounding.

In some disciplines, confounding is categorized into different types. In epidemiology, one type is "confounding by indication", which relates to confounding from observational studies. Because prognostic factors may influence treatment decisions (and bias estimates of treatment effects), controlling for known prognostic factors may reduce this problem, but it is always possible that a forgotten or unknown factor was not included or that factors interact complexly. Confounding by indication has been described as the most important limitation of observational studies. Randomized trials are not affected by confounding by indication due to random assignment.

Confounding variables may also be categorised according to their source. The choice of measurement instrument (operational confound), situational characteristics (procedural confound), or inter-individual differences (person confound).

- An operational confounding can occur in both experimental and non-experimental research designs. This type of confounding occurs when a measure designed to assess a particular construct inadvertently measures something else as well.
- A procedural confounding can occur in a laboratory experiment or a quasi-experiment. This type of confound occurs when the researcher mistakenly allows another variable to change along with the manipulated independent variable.

- A person confounding occurs when two or more groups of units are analyzed together (e.g., workers from different occupations), despite varying according to one or more other (observed or unobserved) characteristics (e.g., gender).

Examples

Say one is studying the relation between birth order (1st child, 2nd child, etc.) and the presence of Down Syndrome in the child. In this scenario, maternal age would be a confounding variable:

- Higher maternal age is directly associated with Down Syndrome in the child.
- Higher maternal age is directly associated with Down Syndrome, regardless of birth order (a mother having her 1st vs 3rd child at age 50 confers the same risk).
- Maternal age is directly associated with birth order (the 2nd child, except in the case of twins, is born when the mother is older than she was for the birth of the 1st child).
- Maternal age is not a consequence of birth order (having a 2nd child does not change the mother's age).

In risk assessments, factors such as age, gender, and educational levels often affect health status and so should be controlled. Beyond these factors, researchers may not consider or have access to data on other causal factors. An example is on the study of smoking tobacco on human health. Smoking, drinking alcohol, and diet are lifestyle activities that are related. A risk assessment that looks at the effects of smoking but does not control for alcohol consumption or diet may overestimate the risk of smoking. Smoking and confounding are reviewed in occupational risk assessments such as the safety of coal mining. When there is not a large sample population of non-smokers or non-drinkers in a particular occupation, the risk assessment may be biased towards finding a negative effect on health.

Decreasing the Potential for Confounding

A reduction in the potential for the occurrence and effect of confounding factors can be obtained by increasing the types and numbers of comparisons performed in an analysis. If measures or manipulations of core constructs are confounded (i.e. operational or procedural confounds exist), subgroup analysis may not reveal problems in the analysis. Additionally, increasing the number of comparisons can create other problems.

Peer review is a process that can assist in reducing instances of confounding, either before study implementation or after analysis has occurred. Peer review relies on collective expertise within a discipline to identify potential weaknesses in study design and analysis, including ways in which results may depend on confounding. Similarly, replication can test for the robustness of findings from one study under alternative study conditions or alternative analyses (e.g., controlling for potential confounds not identified in the initial study).

Confounding effects may be less likely to occur and act similarly at multiple times and locations. In selecting study sites, the environment can be characterized in detail at the study sites to ensure sites are ecologically similar and therefore less likely to have confounding variables. Lastly, the relationship between the environmental variables that possibly confound the analysis and the measured parameters can be studied. The information pertaining to environmental variables can then be used in site-specific models to identify residual variance that may be due to real effects.

Depending on the type of study design in place, there are various ways to modify that design to actively exclude or control confounding variables:

- Case-Control Studies: Case-Control Studies assign confounders to both groups, cases and controls, equally. For example, if somebody wanted to study the cause of myocardial infarct and thinks that the age is a probable confounding variable, each 67-year-old infarct patient will be matched with a healthy 67-year-old "control" person. In case-control studies, matched variables most often are the age and sex. Drawback: Case-control studies are feasible only when it is easy to find controls, *i.e.* persons whose status vis-à-vis all known potential confounding factors is the same as that of the case's patient: Suppose a case-control study attempts to find the cause of a given disease in a person who is 1) 45 years old, 2) African-American, 3) from Alaska, 4) an avid football player, 5) vegetarian, and 6) working in education. A theoretically perfect control would be a person who, in addition to not having the disease being investigated, matches all these characteristics and has no diseases that the patient does not also have—but finding such a control would be an enormous task.
- **Cohort Studies:** A degree of matching is also possible and it is often done by only admitting certain age groups or a certain sex into the study population, creating a cohort of people who share similar characteristics and thus all cohorts are comparable in regard to the possible confounding variable. For example, if age and sex are thought to be confounders, only 40 to 50 years old males would be involved in a cohort study that would assess the myocardial infarct risk in cohorts that either are physically active or inactive. Drawback: In cohort

studies, the overexclusion of input data may lead researchers to define too narrowly the set of similarly situated persons for whom they claim the study to be useful, such that other persons to whom the causal relationship does in fact apply may lose the opportunity to benefit from the study's recommendations. Similarly, "over-stratification" of input data within a study may reduce the sample size in a given stratum to the point where generalizations drawn by observing the members of that stratum alone are not statistically significant.

- **Double Blinding:** Conceals from the trial population and the observers the experiment group membership of the participants. By preventing the participants from knowing if they are receiving treatment or not, the placebo effect should be the same for the control and treatment groups. By preventing the observers from knowing of their membership, there should be no bias from researchers treating the groups differently or from interpreting the outcomes differently.
- **Randomized Controlled Trial:** A method where the study population is divided randomly in order to mitigate the chances of self-selection by participants or bias by the study designers. Before the experiment begins, the testers will assign the members of the participant pool to their groups (control, intervention, parallel), using a randomization process such as the use of a random number generator. For example, in a study on the effects of exercise, the conclusions would be less valid if participants were given a choice if they wanted to belong to the control group which would not exercise or the intervention group which would be willing to take part in an exercise program. The study would then capture other variables besides exercise, such as pre-experiment health levels and motivation to adopt healthy activities. From the observer's side, the experimenter may choose candidates who are more likely to show the results the study wants to see or may interpret subjective results (more energetic, positive attitude) in a way favorable to their desires.
- **Stratification:** As in the example above, physical activity is thought to be a behaviour that protects from myocardial infarct; and age is assumed to be a possible confounder. The data sampled is then stratified by age group – this means that the association between activity and infarct would be analyzed per each age group. If the different age groups (or age strata) yield much different risk ratios, age must be viewed as a confounding variable. There exist statistical tools, among them Mantel–Haenszel methods, that account for stratification of data sets.

- Controlling for confounding by measuring the known confounders and including them as covariates is multivariable analysis such as regression analysis. Multivariate analyses reveal much less information about the *strength* or *polarity* of the confounding variable than do stratification methods. For example, if multivariate analysis controls for antidepressant, and it does not stratify antidepressants for TCA and SSRI, then it will ignore that these two classes of antidepressant have *opposite* effects on myocardial infarction, and one is much *stronger* than the other.

All these methods have their drawbacks:

- The best available defense against the possibility of spurious results due to confounding is often to dispense with efforts at stratification and instead conduct a randomized study of a sufficiently large sample taken as a whole, such that all potential confounding variables (known and unknown) will be distributed by chance across all study groups and hence will be uncorrelated with the binary variable for inclusion/exclusion in any group.
- Ethical considerations: In double-blind and randomized controlled trials, participants are not aware that they are recipients of sham treatments and may be denied effective treatments. There is a possibility that patients only agree to invasive surgery (which carry real medical risks) under the understanding that they are receiving treatment. Although this is an ethical concern, it is not a complete account of the situation. For surgeries that are currently being performed regularly, but for which there is no concrete evidence of a genuine effect, there may be ethical issues to continue such surgeries. In such circumstances, many of people are exposed to the real risks of surgery yet these treatments may possibly offer no discernible benefit. Sham-surgery control is a method that may allow medical science to determine whether a surgical procedure is efficacious or not. Given that there are known risks associated with medical operations, it is questionably ethical to allow unverified surgeries to be conducted ad infinitum into the future.

Artifacts

Artifacts are variables that should have been systematically varied, either within or across studies, but that were accidentally held constant. Artifacts are thus threats to external validity. Artifacts are factors that covary with the treatment and the outcome. Campbell and Stanley identify several artifacts. The major threats to internal validity are history, maturation, testing, instrumentation, statistical regression, selection, experimental mortality, and selection-history interactions.

One way to minimize the influence of artifacts is to use a pretest-posttest control group design. Within this design, "groups of people who are initially equivalent (at the pretest phase) are randomly assigned to receive the experimental treatment or a control condition and then assessed again after this differential experience (posttest phase)". Thus, any effects of artifacts are (ideally) equally distributed in participants in both the treatment and control conditions.

4.1.5 Control Variables

A control variable (or scientific constant) in scientific experimentation is an experimental element which is constant (controlled) and unchanged throughout the course of the investigation. Control variables could strongly influence experimental results were they not held constant during the experiment in order to test the relative relationship of the dependent variable (DV) and independent variable (IV). The control variables themselves are not of primary interest to the experimenter.

"Good controls", also known as "confounders" or "deconfounders", are variables which are theorized to be unaffected by the treatment and which are intended to eliminate omitted-variable bias. "Bad controls", on the other hand, are variables that could be affected by the treatment, might contribute to collider bias, and lead to erroneous results.

Usage

A variable in an experiment which is held constant in order to assess the relationship between multiple variables, is a control variable. A control variable is an element that is not changed throughout an experiment because its unchanging state allows better understanding of the relationship between the other variables being tested.

In any system existing in a natural state, many variables may be interdependent, with each affecting the other. Scientific experiments test the relationship of an IV (or independent variable: that element that is manipulated by the experimenter) to the DV (or dependent variable: that element affected by the manipulation of the IV). Any additional independent variable can be a control variable.

A control variable is an experimental condition or element that is kept the same throughout the experiment, and it is not of primary concern in the experiment, nor will it influence the outcome of the experiment. Any unexpected (e.g.: uncontrolled) change in a control variable during an experiment would invalidate the correlation of dependent variables (DV) to the independent variable (IV), thus skewing the results, and invalidating the working hypothesis. This indicates the presence of a spurious relationship existing within experimental parameters. Unexpected results may result from the presence of a confounding variable, thus requiring a re-working of the initial experimental hypothesis. Confounding variables are a threat to the internal validity of an experiment. This situation may be resolved by first identifying the confounding variable and then redesigning the experiment taking that information into consideration. One way to this is to control the confounding variable, thus making it a control variable. If, however, the spurious relationship cannot be identified, the working hypothesis may have to be abandoned.

Experimental Examples

Take, for example, the well known combined gas law, which is stated mathematically as:

$$\frac{PV}{T} = k$$

where:

- P is the pressure.
- V is the volume.
- T is the thermodynamic temperature measured in kelvins.
- k is a constant (with units of energy divided by temperature).

Which shows that the ratio between the pressure-volume product and the temperature of a system remains constant.

In an experimental verification of parts of the combined gas law, ($P * V = T$), where Pressure, Temperature, and Volume are all variables, to test the resultant changes to any of these variables requires at least one to be kept constant. This is in order to see *comparable experimental results* in the remaining variables.

If Temperature is made the control variable and it is not allowed to change throughout the course of the experiment, the relationship between the dependent variables, Pressure, and Volume, can quickly be established by changing the value for one or the other, and this is Boyle's law. For instance, if the Pressure is raised then the Volume must decrease.

If, however, Volume is made the control variable and it is not allowed to change throughout the course of the experiment, the relationship between dependent variables, Pressure, and Temperature, can quickly be established by changing the value for one or the other, and this is Gay-Lussac's Law. For instance, if the Pressure is raised then the Temperature must increase.

4.1.6 Moderating Variables

In statistics and regression analysis, moderation (also known as effect modification) occurs when the relationship between two variables depends on a third variable. The third variable is referred to as the moderator variable (or effect modifier) or simply the moderator (or modifier). The effect of a moderating variable is characterized statistically as an interaction; that is, a categorical (e.g., sex, ethnicity, class) or continuous (e.g., age, level of reward) variable that is associated with the direction and/or magnitude of the relation between dependent and independent variables. Specifically within a correlational analysis framework, a moderator is a third variable that affects the zero-order correlation between two other variables, or the value of the slope of the dependent variable on the independent variable. In analysis of variance (ANOVA) terms, a basic moderator effect can be represented as an interaction between a focal independent variable and a factor that specifies the appropriate conditions for its operation.

Multicollinearity in Moderated Regression

In moderated regression analysis, a new interaction predictor (x_1x_2) is calculated. However, the new interaction term may be correlated with the two main effects terms used to calculate it. This is the problem of multicollinearity in moderated regression. Multicollinearity tends to cause coefficients to be estimated with higher standard errors and hence greater uncertainty.

Mean-centering (subtracting raw scores from the mean) may reduce multicollinearity, resulting in more interpretable regression coefficients. However, it does not affect the overall model fit.

Post-Hoc Probing of Interactions

Like simple main effect analysis in ANOVA, in post-hoc probing of interactions in regression, we are examining the simple slope of one independent variable at the specific values of the other independent variable. Below is an example of probing two-way interactions. In what follows the regression equation with two variables A and B and an interaction term A*B, $Y = b_0 + b_1A + b_2B + b_3A * B + \varepsilon$ will be considered.

Two Categorical Independent Variables

If both of the independent variables are categorical variables, we can analyze the results of the regression for one independent variable at a specific level of the other independent variable. For example, suppose that both A and B are single dummy coded (0,1) variables, and that A represents ethnicity (0 = European Americans, 1 = East Asians) and B represents the condition in the study (0 = control, 1 = experimental). Then the interaction effect shows whether the effect of condition on the dependent variable Y is different for European Americans and East Asians and whether the effect of ethnic status is different for the two conditions. The coefficient of A shows the ethnicity effect on Y for the control condition, while the coefficient of B shows the effect of imposing the experimental condition for European American participants.

To probe if there is any significant difference between European Americans and East Asians in the experimental condition, we can simply run the analysis with the condition variable reverse-coded (0 = experimental, 1 = control), so that the coefficient for ethnicity represents the ethnicity effect on Y in the experimental condition. In a similar vein, if we want to see whether the treatment has an effect for East Asian participants, we can reverse code the ethnicity variable (0 = East Asians, 1 = European Americans).

One Categorical and One Continuous Independent Variable

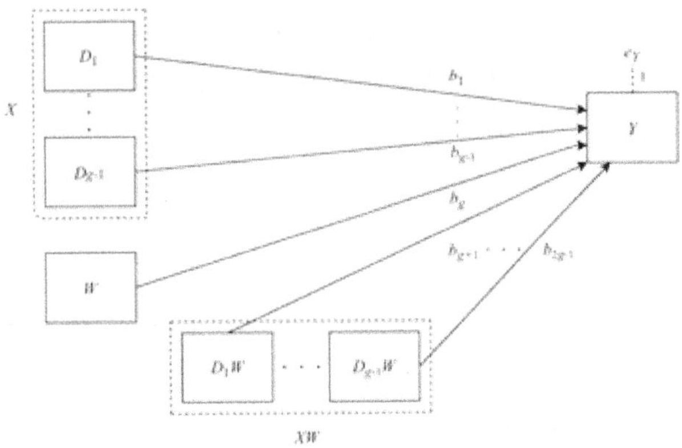

Fig 3: A Statistical Diagram that Depicts a Moderation Model with *X* as a Multicategorical Independent Variable

If the first independent variable is a categorical variable (e.g. gender) and the second is a continuous variable (e.g. scores on the Satisfaction With Life Scale (SWLS)), then b_1 represents the difference in the dependent variable between males and females when life satisfaction is zero.

However, a zero score on the Satisfaction With Life Scale is meaningless as the range of the score is from 7 to 35. This is where centering comes in. If we subtract the mean of the SWLS score for the sample from each participant's score, the mean of the resulting centered SWLS score is zero. When the analysis is run again, b_1 now represents the difference between males and females at the mean level of the SWLS score of the sample.

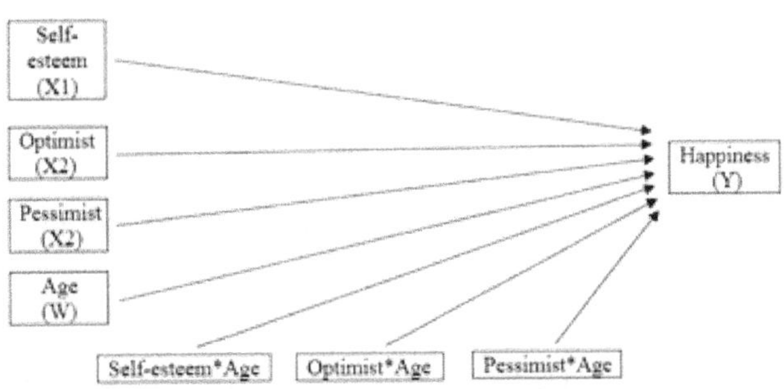

Fig 4: An Example of Conceptual Moderation Model with One Categorical and One Continuous Independent Variable

Cohen et al. (2003) recommended using the following to probe the simple effect of gender on the dependent variable (Y) at three levels of the continuous independent variable: high (one standard deviation above the mean), moderate (at the mean), and low (one standard deviation below the mean).

If the scores of the continuous variable are not standardized, one can just calculate these three values by adding or subtracting one standard deviation of the original scores; if the scores of the continuous variable are standardized, one can calculate the three values as follows: high = the standardized score minus 1, moderate (mean = 0), low = the standardized score plus 1.

Then one can explore the effects of gender on the dependent variable (Y) at high, moderate, and low levels of the SWLS score. As with two categorical independent variables, b_2 represents the effect of the SWLS score on the dependent variable for females.

By reverse coding the gender variable, one can get the effect of the SWLS score on the dependent variable for males.

Coding in Moderated Regression

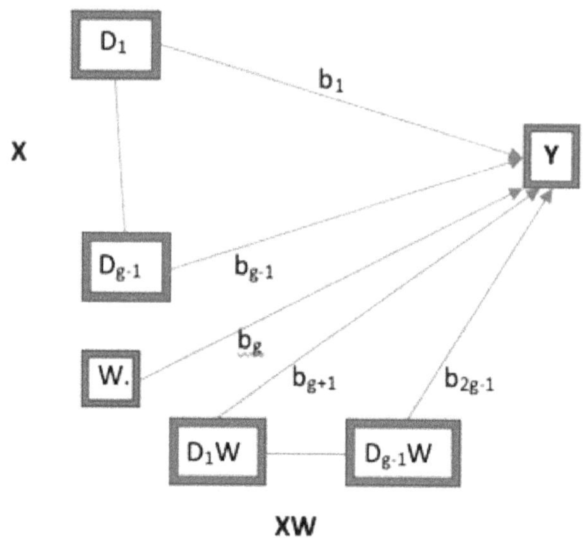

Fig 5: A Statistical Diagram that Depicts a Moderation Model with *W* with Three Levels, as a Multi-Categorical Independent Variable

When treating categorical variables such as ethnic groups and experimental treatments as independent variables in moderated regression, one needs to code the variables so that each code variable represents a specific setting of the categorical variable. There are three basic ways of coding: dummy-variable coding, effects coding, and contrast coding. Below is an introduction to these coding systems.

Groups	D1	D2
No Therapy	-1	0
Cognitive Therapy	0.5	-1
Family Therapy	0.5	1

Dummy coding is used when one has a reference group or one condition in particular (e.g. a control group in the experiment) that is to be compared to each of the other experimental groups. In this case, the intercept is the mean of the reference group, and each of the unstandardized regression coefficients is the difference in the dependent variable between one of the treatment groups and the mean of the reference group (or control group). This coding system is similar to ANOVA analysis, and is appropriate when researchers have a specific reference group and want to compare each of the other groups with it.

Effects coding is used when one does not have a particular comparison or control group and does not have any planned orthogonal contrasts. The intercept is the grand mean (the mean of all the conditions). The regression coefficient is the difference between the mean of one group and the mean of all the group means (e.g. the mean of group A minus the mean of all groups). This coding system is appropriate when the groups represent natural categories.

Contrast coding is used when one has a series of orthogonal contrasts or group comparisons that are to be investigated. In this case, the intercept is the unweighted mean of the individual group means.

The unstandardized regression coefficient represents the difference between the unweighted mean of the means of one group (A) and the unweighted mean of another group (B), where A and B are two sets of groups in the contrast. This coding system is appropriate when researchers have an a priori hypothesis concerning the specific differences among the group means.

Two Continuous Independent Variables

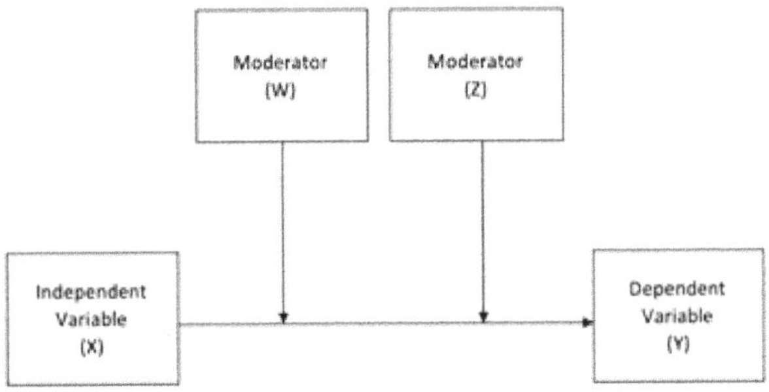

Fig 6: A Conceptual Diagram of an Additive Multiple Moderation Model

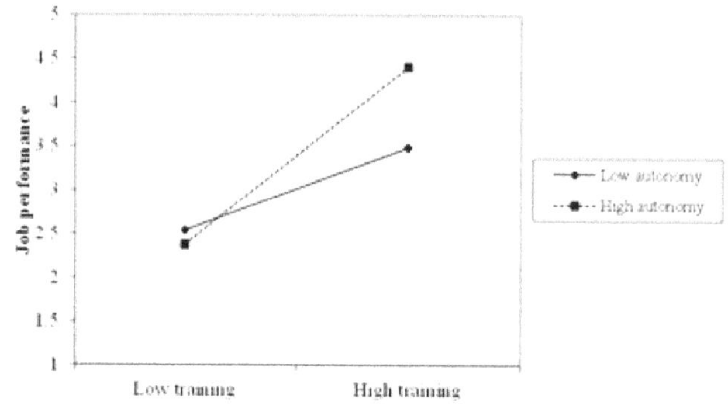

Fig 7: An Example of a Two-Way Interaction Effect Plot

If both of the independent variables are continuous, it is helpful for interpretation to either center or standardize the independent variables, X and Z. (Centering involves subtracting the overall sample mean score from the original score; standardizing does the same followed by dividing by the overall sample standard deviation.)

By centering or standardizing the independent variables, the coefficient of X or Z can be interpreted as the effect of that variable on Y at the mean level of the other independent variable.

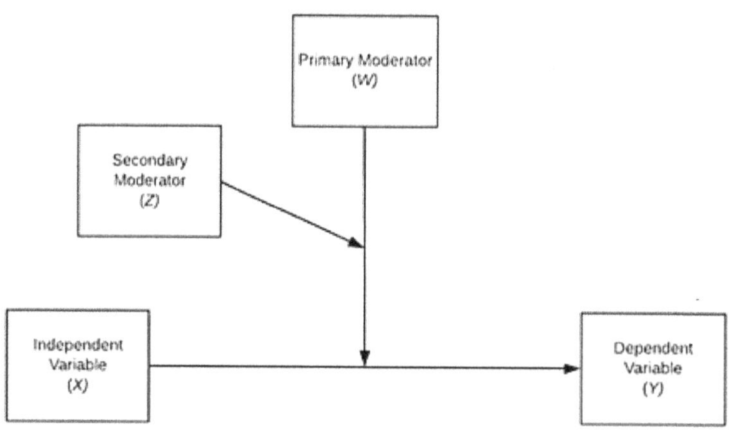

Fig 8: A Conceptual Diagram of a Moderated Moderation Model, otherwise known as a Three-Way Interaction

To probe the interaction effect, it is often helpful to plot the effect of *X* on *Y* at low and high values of *Z* (some people prefer to also plot the effect at moderate values of *Z*, but this is not necessary). Often values of *Z* that are one standard deviation above and below the mean are chosen for this, but any sensible values can be used (and in some cases there are more meaningful values to choose). The plot is usually drawn by evaluating the values of Y for high and low values of both X and Z, and creating two lines to represent the effect of X on Y at the two values of Z. Sometimes this is supplemented by simple slope analysis, which determines whether the effect of X on Y is statistically significant at particular values of Z. A common technique for simple slope analysis is the Johnson-Neyman approach. Various internet-based tools exist to help researchers plot and interpret such two-way interactions.

Higher-Level Interactions

The principles for two-way interactions apply when we want to explore three-way or higher-level interactions. For instance, if we have a three-way interaction between A, B, and C, the regression equation will be as follows:

$$Y = b_0 + b_1 A + b_2 B + b_3 C + b_4 A \cdot B + b_5 A \cdot C + b_6 B \cdot C + b_7 A \cdot B \cdot C + \varepsilon.$$

Spurious Higher-Order Effects

It is worth noting that the reliability of the higher-order terms depends on the reliability of the lower-order terms. For example, if the reliability for variable A is 0.70, the reliability for variable B is 0.80, and their correlation is $r = 0.2$, then the reliability for the interaction variable $A * B$ is $((0.7 \times 0.8) + 0.2^2) / (1 + 0.2^2) = 0.577$. In this case, low reliability of the interaction term leads to low power; therefore, we may not be able to find the interaction effects between A and B that actually exist. The solution for this problem is to use highly reliable measures for each independent variable.

Another caveat for interpreting the interaction effects is that when variable A and variable B are highly correlated, then the $A * B$ term will be highly correlated with the omitted variable A^2; consequently what appears to be a significant moderation effect might actually be a significant nonlinear effect of A alone. If this is the case, it is worth testing a nonlinear regression model by adding nonlinear terms in individual variables into the moderated regression analysis to see if the interactions remain significant. If the interaction effect $A*B$ is still significant, we will be more confident in saying that there is indeed a moderation effect; however, if the interaction effect is no longer significant after adding the nonlinear term, we will be less certain about the existence of a moderation effect and the nonlinear model will be preferred because it is more parsimonious.

Moderated regression analyses also tend to include additional variables, which are conceptualized as covariates of no interest. However, the presence of these covariates can induce spurious effects when either (1) the covariate (C) is correlated with one of the primary variables of interest (e.g. variable A or B), or (2) when the covariate itself is a moderator of the correlation between either A or B with Y. The solution is to include additional interaction terms in the model, for the interaction between each confounder and the primary variables as follows:

$$Y = b_0 + b_1 A + b_2 B + b_3 C + b_4 A \cdot B + b_5 A \cdot C + b_6 B \cdot C + \varepsilon$$

4.1.7 Mediating Variables

In statistics, a mediation model seeks to identify and explain the mechanism or process that underlies an observed relationship between an independent variable and a dependent variable via the inclusion of a third hypothetical variable, known as a mediator variable (also a mediating variable, intermediary variable, or intervening variable).

Rather than a direct causal relationship between the independent variable and the dependent variable, which is often false, a mediation model proposes that the independent variable influences the mediator variable, which in turn influences the dependent variable.

Thus, the mediator variable serves to clarify the nature of the relationship between the independent and dependent variables.

Mediation analyses are employed to understand a known relationship by exploring the underlying mechanism or process by which one variable influences another variable through a mediator variable. In particular, mediation analysis can contribute to better understanding the relationship between an independent variable and a dependent variable when these variables do not have an obvious direct connection.

Direct versus Indirect Effects

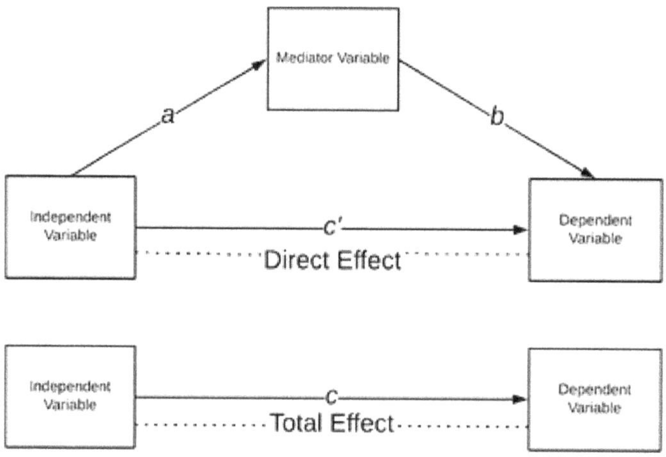

Fig 9: Direct Effect in a Mediation Model

In the diagram shown above, the indirect effect is the product of path coefficients "A" and "B". The direct effect is the coefficient "C".

The direct effect measures the extent to which the dependent variable changes when the independent variable increases by one unit and the mediator variable remains unaltered. In contrast, the indirect effect measures the extent to which the dependent variable changes when the independent variable is held constant and the mediator variable changes by the amount it would have changed had the independent variable increased by one unit.

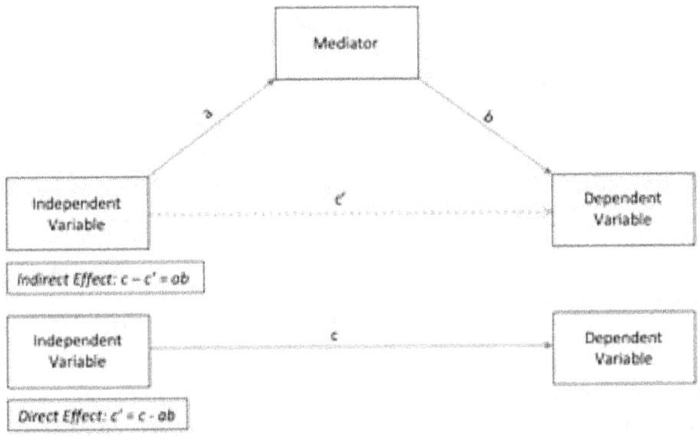

Fig 10: Indirect Effect in a Simple Mediation Model: The Indirect Effect Constitutes the Extent to which the *X* Variable Influences the *Y* Variable through the Mediator

In linear systems, the total effect is equal to the sum of the direct and indirect ($C' + AB$ in the model above). In nonlinear models, the total effect is not generally equal to the sum of the direct and indirect effects, but to a modified combination of the two.

Full Mediation versus Partial Mediation

A mediator variable can either account for all or some of the observed relationship between two variables.

Full Mediation

Maximum evidence for mediation, also called full mediation, would occur if inclusion of the mediation variable drops the relationship between the independent variable and dependent variable to zero.

Partial Mediation

Partial mediation maintains that the mediating variable accounts for some, but not all, of the relationship between the independent variable and dependent variable. Partial mediation implies that there is not only a significant relationship between the mediator and the dependent variable, but also some direct relationship between the independent and dependent variable.

In order for either full or partial mediation to be established, the reduction in variance explained by the independent variable must be significant as determined by one of several tests, such as the Sobel test. The effect of an independent variable on the dependent variable can become nonsignificant when the mediator is introduced simply because a trivial amount of variance is explained (i.e., not true mediation). Thus, it is imperative to show a significant reduction in variance explained by the independent variable before asserting either full or partial mediation. It is possible to have statistically significant indirect effects in the absence of a total effect. This can be explained by the presence of several mediating paths that cancel each other out, and become noticeable when one of the cancelling mediators is controlled for.

This implies that the terms 'partial' and 'full' mediation should always be interpreted relative to the set of variables that are present in the model. In all cases, the operation of "fixing a variable" must be distinguished from that of "controlling for a variable," which has been inappropriately used in the literature. The former stands for physically fixing, while the latter stands for conditioning on, adjusting for, or adding to the regression model.

The two notions coincide only when all error terms (not shown in the diagram) are statistically uncorrelated. When errors are correlated, adjustments must be made to neutralize those correlations before embarking on mediation analysis.

Sobel's Test

Sobel's test is performed to determine if the relationship between the independent variable and dependent variable has been significantly reduced after inclusion of the mediator variable. In other words, this test assesses whether a mediation effect is significant.

It examines the relationship between the independent variable and the dependent variable compared to the relationship between the independent variable and dependent variable including the mediation factor.

The Sobel test is more accurate than the Baron and Kenny steps explained above; however, it does have low statistical power. As such, large sample sizes are required in order to have sufficient power to detect significant effects. This is because the key assumption of Sobel's test is the assumption of normality. Because Sobel's test evaluates a given sample on the normal distribution, small sample sizes and skewness of the sampling distribution can be problematic.

Thus, the rule of thumb as suggested by MacKinnon et al., (2002) is that a sample size of 1000 is required to detect a small effect, a sample size of 100 is sufficient in detecting a medium effect, and a sample size of 50 is required to detect a large effect.

The equation for Sobel is:

$$z = \frac{ab}{\sqrt{b^2 s_a^2 + a^2 s_b^2}}$$

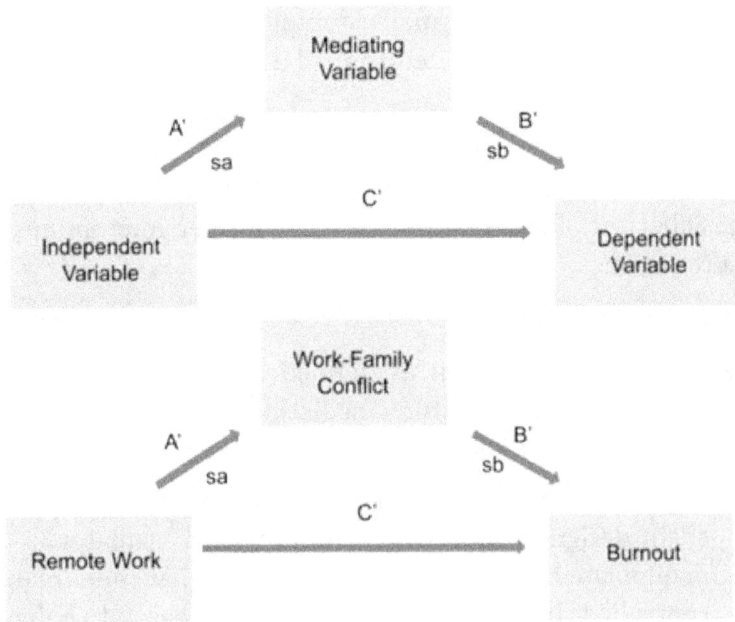

Fig 11: Sobel Test where the IV is Remote Work

Preacher–Hayes Bootstrap Method

The bootstrapping method provides some advantages to the Sobel's test, primarily an increase in power. The Preacher and Hayes bootstrapping method is a non-parametric test and does not impose the assumption of normality. Therefore, if the raw data is available, the bootstrap method is recommended. Bootstrapping involves repeatedly randomly sampling observations with replacement from the data set to compute the desired statistic in each resample. Computing over hundreds, or thousands, of bootstrap resamples provide an approximation of the sampling distribution of the statistic of interest.

The Preacher–Hayes method provides point estimates and confidence intervals by which one can assess the significance or nonsignificance of a mediation effect. Point estimates reveal the mean over the number of bootstrapped samples and if zero does not fall between the resulting confidence intervals of the bootstrapping method, one can confidently conclude that there is a significant mediation effect to report.

Significance of Mediation

There are a few different options one can choose from to evaluate a mediation model. Bootstrapping is becoming the most popular method of testing mediation because it does not require the normality assumption to be met, and because it can be effectively utilized with smaller sample sizes ($N < 25$). However, mediation continues to be most frequently determined using the logic of Baron and Kenny or the Sobel test.

It is becoming increasingly more difficult to publish tests of mediation based purely on the Baron and Kenny method or tests that make distributional assumptions such as the Sobel test. Thus, it is important to consider your options when choosing which test to conduct.

Approaches to Mediation

While the concept of mediation as defined within psychology is theoretically appealing, the methods used to study mediation empirically have been challenged by statisticians and epidemiologists and interpreted formally.

- **Experimental-Causal-Chain Design:** An experimental-causal-chain design is used when the proposed mediator is experimentally manipulated. Such a design implies that one manipulates some controlled third variable that they have reason to believe could be the underlying mechanism of a given relationship.
- **Measurement-of-Mediation Design:** A measurement-of-mediation design can be conceptualized as a statistical approach. Such a design implies that one measures the proposed intervening variable and then uses statistical analyses to establish mediation. This approach does not involve manipulation of the hypothesized mediating variable, but only involves measurement.

Criticisms of Mediation Measurement

Potentially Unnecessary Step

Hayes (2009) critiqued Baron and Kenny's mediation steps approach, and as of 2019, David A. Kenny on his website stated that mediation can exist in the absence of a 'significant' total effect (sometimes referred to as "inconsistent mediation"), and therefore step 1 of the original 1986 approach may not be needed.

Later publications by Hayes questioned the concepts of full mediation and partial mediation, and advocated for the abandonment of these terms and of the steps in classical (1986) mediation.

Importance of Caution

Experimental approaches to mediation must be carried out with caution. First, it is important to have strong theoretical support for the exploratory investigation of a potential mediating variable.

A criticism of a mediation approach rests on the ability to manipulate and measure a mediating variable. Thus, one must be able to manipulate the proposed mediator in an acceptable and ethical fashion.

As such, one must be able to measure the intervening process without interfering with the outcome. The mediator must also be able to establish construct validity of manipulation.

One of the most common criticisms of the measurement-of-mediation approach is that it is ultimately a correlational design. Consequently, it is possible that some other third variable, independent from the proposed mediator, could be responsible for the proposed effect. However, researchers have worked hard to provide counter-evidence to this disparagement. Specifically, the following counter-arguments have been put forward:

- **Temporal Precedence:** For example, if the independent variable precedes the dependent variable in time, this would provide evidence suggesting a directional, and potentially causal, link from the independent variable to the dependent variable.
- **Nonspuriousness and/or No Confounds:** For example, should one identify other third variables and prove that they do not alter the relationship between the independent variable and the dependent variable he/she would have a stronger argument for their mediation effect.

Mediation can be an extremely useful and powerful statistical test; however, it must be used properly. It is important that the measures used to assess the mediator and the dependent variable are theoretically distinct and that the independent variable and mediator cannot interact.

Should there be an interaction between the independent variable and the mediator one would have grounds to investigate moderation.

Other Third Variables

Confounding

Another model that is often tested is one in which competing variables in the model are alternative potential mediators or an unmeasured cause of the dependent variable. An additional variable in a causal model may obscure or confound the relationship between the independent and dependent variables.

Potential confounders are variables that may have a causal impact on both the independent variable and dependent variable. They include common sources of measurement error as well as other influences shared by both the independent and dependent variables.

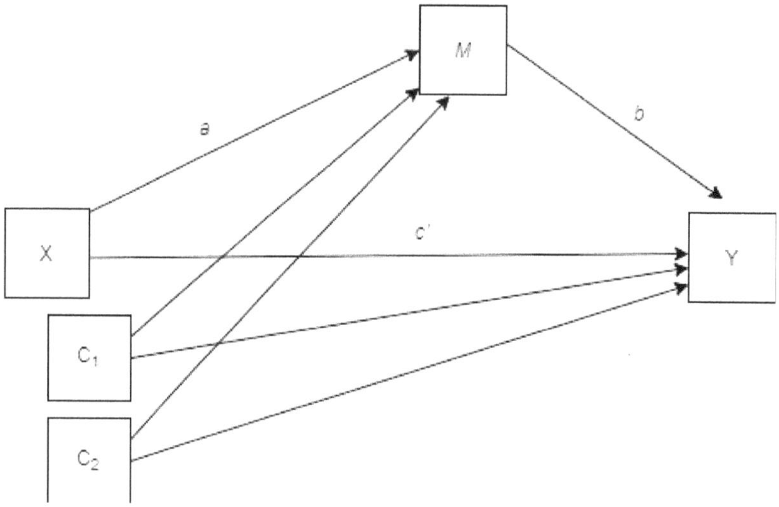

Fig 12: Mediation Model with Two Covariates

In experimental studies, there is a special concern about aspects of the experimental manipulation or setting that may account for study effects, rather than the motivating theoretical factor. Any of these problems may produce spurious relationships between the independent and dependent variables as measured. Ignoring a confounding variable may bias empirical estimates of the causal effect of the independent variable.

Suppression

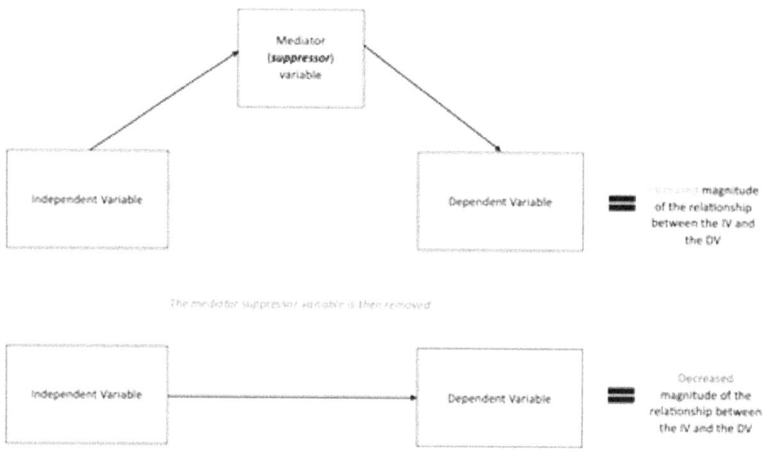

Fig 13: A Moderator Variable that Increases the Predictive Validity of Another Variable is known as a Suppression Variable

A suppressor variable increases the predictive validity of another variable when included in a regression equation. Suppression can occur when a single causal variable is related to an outcome variable through two separate mediator variables, and when one of those mediated effects is positive and one is negative. In such a case, each mediator variable suppresses or conceals the effect that is carried through the other mediator variable.

For example, higher intelligence scores (a causal variable, *A*) may cause an increase in error detection (a mediator variable, *B*) which in turn may cause a decrease in errors made at work on an assembly line (an outcome variable, *X*); at the same time, intelligence could also cause an increase in boredom (*C*), which in turn may cause an *increase* in errors (*X*).

Thus, in one causal path intelligence decreases errors, and in the other it increases them. When neither mediator is included in the analysis, intelligence appears to have no effect or a weak effect on errors. However, when boredom is controlled intelligence will appear to decrease errors, and when error detection is controlled intelligence will appear to increase errors. If intelligence could be increased while only boredom was held constant, errors would decrease; if intelligence could be increased while holding only error detection constant, errors would increase. In general, the omission of suppressors or confounders will lead to either an underestimation or an overestimation of the effect of *A* on X, thereby either reducing or artificially inflating the magnitude of a relationship between two variables.

Moderators

Other important third variables are moderators. Moderators are variables that can make the relationship between two variables either stronger or weaker. Such variables further characterize interactions in regression by affecting the direction and/or strength of the relationship between *X* and *Y*. A moderating relationship can be thought of as an interaction. It occurs when the relationship between variables A and B depends on the level of C.

Moderated Mediation

Mediation and moderation can co-occur in statistical models. It is possible to mediate moderation and moderate mediation. Moderated mediation is when the effect of the treatment *A* on the mediator and/or the partial effect *B* on the dependent variable depend in turn on levels of another variable (moderator). Essentially, in moderated mediation, mediation is first established, and then one investigates if the mediation effect that describes the relationship between the independent variable and dependent variable is moderated by different levels of another variable (i.e., a moderator). This definition has been outlined by Muller, Judd, and Yzerbyt (2005) and Preacher, Rucker, and Hayes (2007).

Models of Moderated Mediation

There are five possible models of moderated mediation, as illustrated in the diagrams below.

- In the first model the independent variable also moderates the relationship between the mediator and the dependent variable.
- The second possible model of moderated mediation involves a new variable which moderates the relationship between the independent variable and the mediator (the *A* path).
- The third model of moderated mediation involves a new moderator variable which moderates the relationship between the mediator and the dependent variable (the *B* path).
- Moderated mediation can also occur when one moderating variable affects both the relationship between the independent variable and the mediator (the *A* path) and the relationship between the mediator and the dependent variable (the *B* path).
- The fifth and final possible model of moderated mediation involves two new moderator variables, one moderating the *A* path and the other moderating the *B* path.

In addition to the models mentioned above, a new variable can also exist which moderates the relationship between the independent variable and mediator (the A path) while at the same time have the new variable moderate the relationship between the independent variable and dependent variable (the C Path).

Mediated Moderation

Mediated moderation is a variant of both moderation and mediation. This is where there is initially overall moderation and the direct effect of the moderator variable on the outcome is mediated. The main difference between mediated moderation and moderated mediation is that for the former there is initial (overall) moderation and this effect is mediated and for the latter there is no moderation but the effect of either the treatment on the mediator (path *A*) is moderated or the effect of the mediator on the outcome (path *B*) is moderated.

In order to establish mediated moderation, one must first establish moderation, meaning that the direction and/or the strength of the relationship between the independent and dependent variables (path *C*) differs depending on the level of a third variable (the moderator variable). Researchers next look for the presence of mediated moderation when they have a theoretical reason to believe that there is a fourth variable that acts as the mechanism or process that causes the relationship between the independent variable and the moderator (path *A*) or between the moderator and the dependent variable (path *C*).

Example

The following is a published example of mediated moderation in psychological research. Participants were presented with an initial stimulus (a prime) that made them think of morality or made them think of might. They then participated in the Prisoner's Dilemma Game (PDG), in which participants pretend that they and their partner in crime have been arrested, and they must decide whether to remain loyal to their partner or to compete with their partner and cooperate with the authorities. The researchers found that prosocial individuals were affected by the morality and might primes, whereas proself individuals were not. Thus, social value orientation (proself vs. prosocial) moderated the relationship between the prime (independent variable: morality vs. might) and the behaviour chosen in the PDG (dependent variable: competitive vs. cooperative). The researchers next looked for the presence of a mediated moderation effect. Regression analyses revealed that the type of prime (morality vs. might) mediated the moderating relationship of participants' social value orientation on PDG behaviour. Prosocial participants who experienced the morality prime expected their partner to cooperate with them, so they chose to cooperate themselves. Prosocial participants who experienced the might prime expected their partner to compete with them, which made them more likely to compete with their partner and cooperate with the authorities. In contrast, participants with a pro-self social value orientation always acted competitively.

4.2 ROLE OF VARIABLES IN BIOSTATISTICS

Biostatistics is the application of statistical techniques to biological and health-related data. Variables are central to biostatistics, as they represent the data that is collected, analyzed, and interpreted. The types of variables in a study determine the choice of statistical methods and the validity of the results. Key areas in which variables are crucial in biostatistics include:

4.2.1 Descriptive Statistics

Descriptive statistics provide a summary of the data in terms of measures such as mean, median, mode, standard deviation, and range. These statistics describe the characteristics of the sample and help researchers understand the distribution of the variables being studied. Descriptive statistics are used to summarize both continuous and categorical variables. For continuous variables, measures of central tendency (mean, median, mode) and variability (standard deviation, variance) are typically reported. For categorical variables, frequencies and percentages are used to describe the distribution of different categories.

4.2.2 Inferential Statistics

Inferential statistics allow researchers to make generalizations or predictions about a population based on a sample of data. The choice of inferential statistical methods depends on the type of variables involved. Common inferential techniques include:

- **T-tests and ANOVA:** These tests are used to compare means between groups for continuous variables. A t-test compares the means of two groups, while ANOVA compares the means of three or more groups.
- **Chi-Square Test:** The chi-square test is used to determine whether there is an association between two categorical variables. It compares the observed frequencies of categorical data with the expected frequencies under the assumption of no association.
- **Regression Analysis:** Regression analysis is used to model the relationship between an independent variable and a dependent variable. Linear regression is used when the dependent variable is continuous, while logistic regression is used when the dependent variable is binary (e.g., disease/no disease).

4.2.3 Hypothesis Testing

Variables are central to hypothesis testing, a key aspect of research methods in biostatistics. In hypothesis testing, researchers begin with a null hypothesis, which assumes no effect or relationship between the variables being studied. Statistical tests are then applied to determine whether the observed data provide sufficient evidence to reject the null hypothesis in favor of an alternative hypothesis. The choice of hypothesis test depends on the type of variables involved. For example, if researchers want to test whether there is a difference in blood pressure between individuals receiving two different treatments, a t-test for independent samples might be used if the dependent variable (blood pressure) is continuous. In contrast, if the research question involves comparing proportions between groups (e.g., the proportion of smokers vs. non-smokers with lung disease), a chi-square test may be appropriate.

4.3 VARIABLES IN RESEARCH DESIGN

Variables also play a pivotal role in the design of research studies. The selection and operationalization of variables are critical steps in developing a valid study design. Researchers must define their independent and dependent variables clearly and consider potential confounding variables that may affect the results.

4.3.1 Operationalization of Variables

Operationalization involves defining how variables will be measured or manipulated in a study. For continuous variables such as blood pressure, this might involve specifying the device and method used for measurement. For categorical variables, operationalization involves defining the categories (e.g., smoker vs. non-smoker). Clear operational definitions are essential for ensuring that variables are measured consistently across study participants.

4.3.2 Study Design and Variables

Different types of research designs, such as experimental, observational, or cohort studies, involve different uses of variables. In experimental studies, researchers manipulate the independent variable to observe its effect on the dependent variable, while controlling for other variables. Observational studies, on the other hand, involve measuring variables without manipulation, and the goal is often to identify associations between variables.

4.4 CONCLUSION

Variables are at the heart of biostatistics and research methods, driving the design, analysis, and interpretation of data. Whether they are continuous or categorical, dependent or independent, variables allow researchers to quantify phenomena, test hypotheses, and draw meaningful conclusions. By understanding the types and roles of variables in research, biostatisticians can apply appropriate statistical methods to advance knowledge in health and medicine. Proper management of variables ensures the validity of study findings and enhances the quality of research outcomes, ultimately contributing to evidence-based practice in healthcare.

CHAPTER 5
FREQUENCY DISTRIBUTION

In biostatistics and research methods, frequency distribution is a crucial tool for organizing and summarizing data. It involves grouping data into categories and showing how frequently each category or value occurs in the dataset. This method provides a snapshot of the data's structure, helping researchers better understand patterns, relationships, and trends. Frequency distribution is essential for both descriptive and inferential statistics, as it forms the foundation for calculating measures of central tendency (like the mean and median) and dispersion (like the range and standard deviation).

5.1 UNDERSTANDING FREQUENCY DISTRIBUTION

5.1.1 Definition

A frequency distribution is a tabular or graphical representation of data that shows the frequency of each value or category. In simpler terms, it answers the question: "How often does each value occur?" The data can be grouped into intervals (also called bins or classes) for continuous variables or listed as categories for discrete variables.

For example, if we are looking at the ages of a group of patients in a study, a frequency distribution would tell us how many patients fall into each age group. This allows researchers to see the spread of the data at a glance.

5.1.2 Components of Frequency Distribution

A frequency distribution typically consists of the following components:

- **Values or Classes:** These represent the different possible values of a variable or ranges into which data is grouped.
- **Frequencies:** These are the counts of how many times each value or class occurs in the dataset.
- **Cumulative Frequency:** This is the running total of frequencies up to a particular class or value, showing how many observations fall below a certain point.
- **Relative Frequency:** The frequency of a class divided by the total number of observations, usually expressed as a proportion or

percentage. Relative frequencies are useful for comparing distributions with different sample sizes.

5.1.3 Types of Frequency Distribution

There are several types of frequency distributions used in biostatistics and research methods, depending on the nature of the data:

- **Simple Frequency Distribution:** This involves listing each distinct value of the variable along with its corresponding frequency. It's useful for small datasets with discrete variables.
- **Grouped Frequency Distribution:** For large datasets or continuous variables, data is grouped into intervals or classes. Each class represents a range of values, and the frequency indicates how many observations fall within that range.
- **Cumulative Frequency Distribution:** This type of distribution shows the total number of observations that fall below the upper boundary of each class, providing insights into the cumulative effect of the data.
- **Relative Frequency Distribution:** Instead of showing the actual counts, a relative frequency distribution displays the proportion of the total number of observations that fall into each category or class.

5.2 CONSTRUCTION OF FREQUENCY DISTRIBUTION

5.2.1 Steps in Constructing a Frequency Distribution

- **Determine the Range of the Data:** First, calculate the range, which is the difference between the maximum and minimum values in the dataset. This will help in determining the intervals or classes for grouping the data.
- **Choose the Number of Classes:** Decide how many intervals or classes to divide the data into. For large datasets, it is common to use between 5 and 20 classes, depending on the data's spread and variability.
- **Determine Class Width:** The width of each class can be calculated by dividing the range of the data by the number of classes. The class width should be consistent across all intervals.
- **Assign Class Intervals:** Once the class width is determined, create the class intervals. Make sure the intervals are mutually exclusive and exhaustive so that every observation falls into exactly one class.
- **Tally the Data:** Go through the dataset and assign each value to its appropriate class. This process is known as tallying, and it helps in counting the frequency for each class.

- **Calculate Frequencies:** After tallying the data, count the number of values in each class or category to obtain the frequencies. You can then create a table with class intervals and their corresponding frequencies.
- **Calculate Cumulative and Relative Frequencies:** If necessary, compute the cumulative and relative frequencies by summing the frequencies progressively and converting the frequencies into proportions or percentages, respectively.

5.2.2 Example of Frequency Distribution

Let's consider an example of a dataset that includes the ages of 50 patients in a hospital. The ages range from 20 to 80 years. By following the steps outlined above, we can create a frequency distribution with five class intervals (e.g., 20-29, 30-39, and so on). We then tally the data, calculate the frequencies, and present the data in a table. This table allows us to quickly see how the patients' ages are distributed across the intervals.

Age Group	Frequency	Cumulative Frequency	Relative Frequency (%)
20-29	5	5	10%
30-39	10	15	20%
40-49	15	30	30%
50-59	12	42	24%
60-69	6	48	12%
70-79	2	50	4%

5.3 GRAPHICAL REPRESENTATION OF FREQUENCY DISTRIBUTION

Visualizing frequency distribution is crucial for understanding data patterns. Several graphical techniques are commonly used in biostatistics:

5.3.1 Histograms

A histogram is a visual representation of the distribution of quantitative data. To construct a histogram, the first step is to "bin" (or "bucket") the range of values—divide the entire range of values into a series of intervals—and then count how many values fall into each interval. The bins are usually specified as consecutive, non-overlapping intervals of a variable. The bins (intervals) are adjacent and are typically (but not required to be) of equal size.

Histograms give a rough sense of the density of the underlying distribution of the data, and often for density estimation: estimating the probability density function of the underlying variable.

The total area of a histogram used for probability density is always normalized to 1. If the length of the intervals on the x-axis are all 1, then a histogram is identical to a relative frequency plot.

Histograms are sometimes confused with bar charts. In a histogram, each bin is for a different range of values, so altogether the histogram illustrates the distribution of values. But in a bar chart, each bar is for a different category of observations (e.g., each bar might be for a different population), so altogether the bar chart can be used to compare different categories. Some authors recommend that bar charts always have gaps between the bars to clarify that they are not histograms.

Mathematical Definitions

The data used to construct a histogram are generated via a function m_i that counts the number of observations that fall into each of the disjoint categories (known as *bins*). Thus, if we let n be the total number of observations and k be the total number of bins, the histogram data m_i meet the following conditions:

$$n = \sum_{i=1}^{k} m_i .$$

A histogram can be thought of as a simplistic kernel density estimation, which uses a kernel to smooth frequencies over the bins. This yields a smoother probability density function, which will in general more accurately reflect distribution of the underlying variable. The density estimate could be plotted as an alternative to the histogram, and is usually drawn as a curve rather than a set of boxes. Histograms are nevertheless preferred in applications, when their statistical properties need to be modeled. The correlated variation of a kernel density estimate is very difficult to describe mathematically, while it is simple for a histogram where each bin varies independently. An alternative to kernel density estimation is the average shifted histogram, which is fast to compute and gives a smooth curve estimate of the density without using kernels.

Cumulative Histogram

A cumulative histogram is a mapping that counts the cumulative number of observations in all of the bins up to the specified bin. That is, the cumulative histogram M_i of a histogram m_j is defined as:

$$M_i = \sum_{j=1}^{i} m_j .$$

Number of Bins and Width

There is no "best" number of bins, and different bin sizes can reveal different features of the data. Grouping data is at least as old as Graunt's work in the 17th century, but no systematic guidelines were given until Sturges's work in 1926.

Using wider bins where the density of the underlying data points is low reduces noise due to sampling randomness; using narrower bins where the density is high (so the signal drowns the noise) gives greater precision to the density estimation. Thus varying the bin-width within a histogram can be beneficial. Nonetheless, equal-width bins are widely used.

Some theoreticians have attempted to determine an optimal number of bins, but these methods generally make strong assumptions about the shape of the distribution. Depending on the actual data distribution and the goals of the analysis, different bin widths may be appropriate, so experimentation is usually needed to determine an appropriate width. There are, however, various useful guidelines and rules of thumb.

The number of bins k can be assigned directly or can be calculated from a suggested bin width h as:

$$k = \left\lceil \frac{\max x - \min x}{h} \right\rceil.$$

The braces indicate the ceiling function.

Applications

- In hydrology the histogram and estimated density function of rainfall and river discharge data, analysed with a probability distribution, are used to gain insight in their behaviour and frequency of occurrence. An example is shown in the blue figure.
- In many Digital image processing programs there is an histogram tool, which show you the distribution of the contrast / brightness of the pixels.

5.3.2 Frequency Polygons

A frequency polygon is a line graph that connects points corresponding to the frequencies of class intervals. The x-axis represents the class midpoints, while the y-axis represents the frequencies. Frequency polygons are useful for comparing multiple datasets and understanding overall trends in the data.

5.3.3 Cumulative Frequency Graphs (Ogives)

In statistics, an ogive, also known as a cumulative frequency polygon, can refer to one of two things:

- Any hand-drawn graphic of a cumulative distribution function.
- Any empirical cumulative distribution function.

The points plotted as part of an ogive are the upper class limit and the corresponding cumulative absolute frequency or cumulative relative frequency. The ogive for the normal distribution (on one side of the mean) resembles (one side of) an Arabesque or ogival arch, which is likely the origin of its name.

Creation

Along the horizontal axis, the limits of the class intervals for an ogive are marked. Based on the limit values, points above each are placed with heights equal to either the absolute or relative cumulative frequency. The shape of an ogive is obtained by connecting each of the points to its neighbours with line segments. Sometimes an axis for both the absolute frequency and relative is drawn.

Finding Percentages

Ogives, similarly to other representations of cumulative distribution functions, are useful for estimating centiles in a distribution. For example, we can know the central point so that 50% of the observations would be below this point and 50% above. To do this, we draw a line from the point of 50% on the axis of percentage until it intersects with the curve. Then we vertically project the intersection onto the horizontal axis. The last intersection gives us the desired value. The frequency polygon and ogive are used to compare two statistical sets whose number could be different.

5.3.4 Pie Charts and Bar Charts

Pie Chart

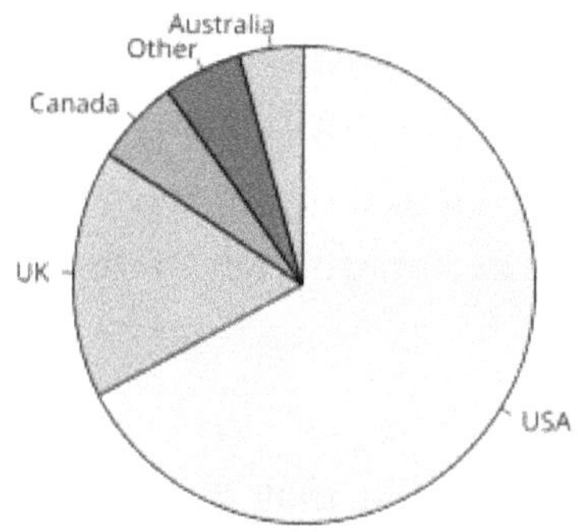

Fig 1: Pie Chart of Populations of English Native Speakers

A pie chart (or a circle chart) is a circular statistical graphic which is divided into slices to illustrate numerical proportion. In a pie chart, the arc length of each slice (and consequently its central angle and area) is proportional to the quantity it represents. While it is named for its resemblance to a pie which has been sliced, there are variations on the way it can be presented. The earliest known pie chart is generally credited to William Playfair's *Statistical Breviary* of 1801.

Pie charts are very widely used in the business world and the mass media. However, they have been criticized, and many experts recommend avoiding them, as research has shown it is difficult to compare different sections of a given pie chart, or to compare data across different pie charts. Pie charts can be replaced in most cases by other plots such as the bar chart, box plot, dot plot, etc.

History

The earliest known pie chart is generally credited to William Playfair's *Statistical Breviary* of 1801, in which two such graphs are used.

Playfair presented an illustration, which contained a series of pie charts. One of those charts depicted the proportions of the Turkish Empire located in Asia, Europe and Africa before 1789. This invention was not widely used at first.

Playfair thought that pie charts were in need of a third dimension to add additional information.

Florence Nightingale may not have invented the pie chart, but she adapted it to make it more readable, which fostered its wide use, still today. Nightingale reconfigured the pie chart making the length of the wedges variable instead of their width.

The graph, then, resembled a cock's comb. She was later assumed to have created it due to the obscurity and lack of practicality of Playfair's creation. Nightingale's polar area diagram, or occasionally the Nightingale rose diagram, equivalent to a modern circular histogram, to illustrate seasonal sources of patient mortality in the military field hospital she managed, was published in Notes on Matters Affecting the Health, Efficiency, and Hospital Administration of the British Army and sent to Queen Victoria in 1858.

According to the historian Hugh Small, "she may have been the first to use [pie charts] for persuading people of the need for change."

3D Pie Chart and Perspective Pie Cake

A 3D pie chart, or perspective pie chart, is used to give the chart a 3D look. Often used for aesthetic reasons, the third dimension does not improve the reading of the data; on the contrary, these plots are difficult to interpret because of the distorted effect of perspective associated with the third dimension.

The use of superfluous dimensions not used to display the data of interest is discouraged for charts in general, not only for pie charts.

Doughnut Chart

A doughnut chart (also spelled donut) is a variant of the pie chart, with a blank center allowing for additional information about the data as a whole to be included.

Doughnut charts are similar to pie charts in that their aim is to illustrate proportions. This type of circular graph can support multiple statistics at once and it provides a better data intensity ratio to standard pie charts. It does not have to contain information in the center.

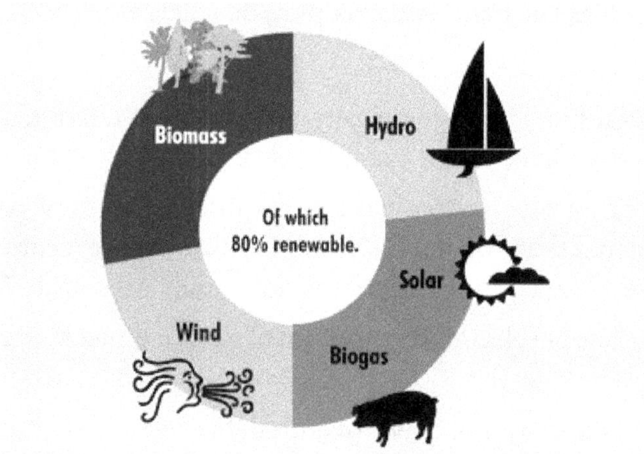

Fig 2: Information about the Data as a Whole in the Center of a Doughnut Chart

Exploded Pie Chart

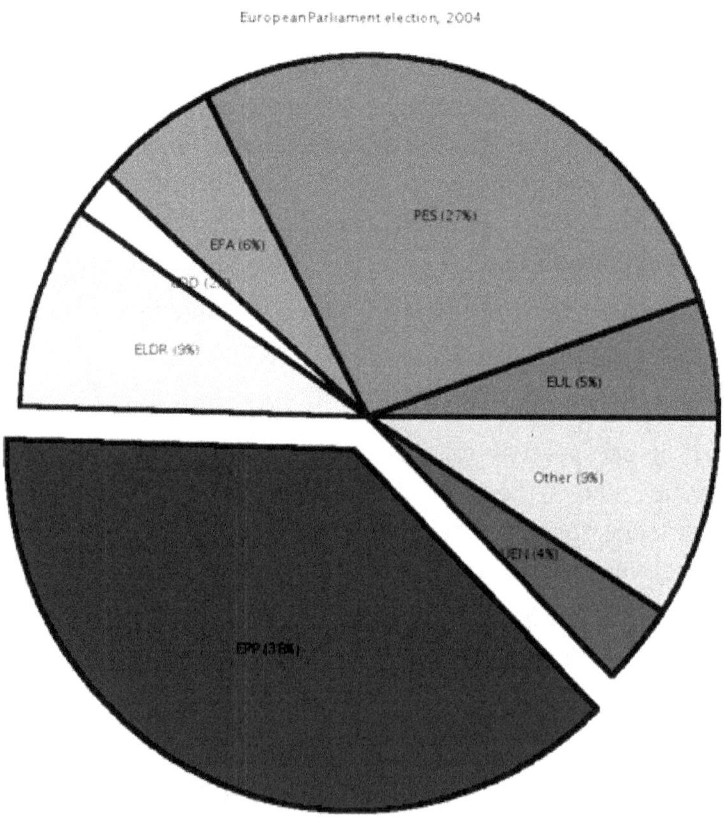

Fig 3: Exploded Pie Chart for the Example Data with the Largest Party Group Exploded

A chart with one or more sectors separated from the rest of the disk is known as an *exploded pie chart*. This effect is used to either highlight a sector, or to highlight smaller segments of the chart with small proportions.

Polar Area Diagram

The polar area diagram is similar to a usual pie chart, except sectors have equal angles and differ rather in how far each sector extends from the center of the circle. The polar area diagram is used to plot cyclic phenomena (e.g., counts of deaths by month). For example, if the counts of deaths in each month for a year are to be plotted then there will be 12 sectors (one per month) all with the same angle of 30 degrees each. The radius of each sector would be proportional to the square root of the death rate for the month, so the area of a sector represents the rate of deaths in a month. If the death rate in each month is subdivided by cause of death, it is possible to make multiple comparisons on one diagram, as is seen in the polar area diagram famously developed by Florence Nightingale.

The first known use of polar area diagrams was by André-Michel Guerry, which he called *courbes circulaires* (circular curves), in an 1829 paper showing seasonal and daily variation in wind direction over the year and births and deaths by hour of the day.

Ring Chart, Sunburst Chart, and Multilevel Pie Chart

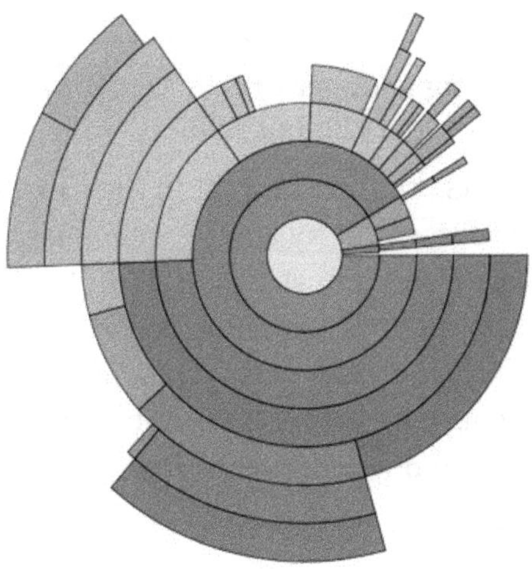

Fig 4: Multi-Level Pie Chart Representing Disk Usage in a Linux File System

Léon Lalanne later used a polar diagram to show the frequency of wind directions around compass points in 1843. The wind rose is still used by meteorologists. Nightingale published her rose diagram in 1858. Although the name "coxcomb" has come to be associated with this type of diagram, Nightingale originally used the term to refer to the publication in which this diagram first appeared—an attention-getting book of charts and tables—rather than to this specific type of diagram. A ring chart, also known as a sunburst chart or a multilevel pie chart, is used to visualize hierarchical data, depicted by concentric circles. The circle in the center represents the root node, with the hierarchy moving outward from the center. A segment of the inner circle bears a hierarchical relationship to those segments of the outer circle which lie within the angular sweep of the parent segment.

Spie Chart

A variant of the polar area chart is the spie chart, designed by Dror Feitelson. The design superimposes a normal pie chart with a modified polar area chart to permit the comparison of two sets of related data.

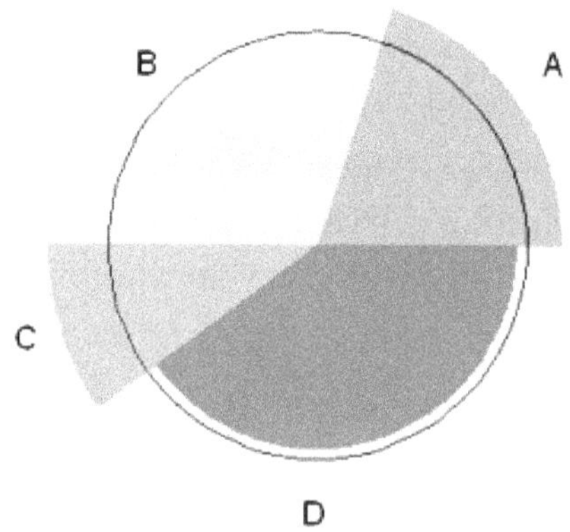

Fig 5: A Spie Chart Comparing Number of Students with Student Costs Across Four Different Schools

The base pie chart represents the first data set in the usual way, with different slice sizes. The second set is represented by the superimposed polar area chart, using the same angles as the base, and adjusting the radii to fit the data. For example, the base pie chart could show the distribution of age and gender groups in a population, and the overlay their representation among road casualties. Age and gender groups that are especially susceptible to being involved in accidents then stand out as slices that extend beyond the original pie chart.

Square Chart/Waffle Chart

Square charts, also called waffle charts, are a form of pie charts that use squares instead of circles to represent percentages. Similar to basic circular pie charts, square pie charts take each percentage out of a total 100%. They are often 10 by 10 grids, where each cell represents 1%. Despite the name, circles, pictograms (such as of people), and other shapes may be used instead of squares. One major benefit to square charts is that smaller percentages, difficult to see on traditional pie charts, can be easily depicted.

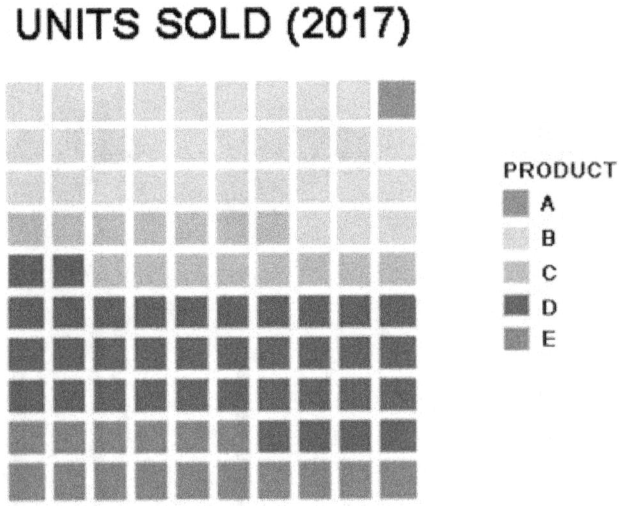

Fig 6: Square Pie Chart (Waffle Chart), Showing how Smaller Percentages are More Easily Shown than on Circular Charts. On the 10x10 Grid, Each Cell Represents 1%

Use and Effectiveness

Atmospheric air content	
Gas	*percentage*
Nitrogen	78.00%
Oxygen	21.00%
Others	1.00%

Fig 7: 3D Pie Chart Showing Atmospheric Air Components Percentage

A flaw exhibited by pie charts is that they cannot show more than a few values without separating the visual encoding (the "slices") from the data they represent (typically percentages). When slices become too small, pie charts have to rely on colors, textures or arrows so the reader can understand them. This makes them unsuitable for use with larger amounts of data. Pie charts also take up a larger amount of space on the page compared to the more flexible bar charts, which do not need to have separate legends, and can display other values such as averages or targets at the same time.

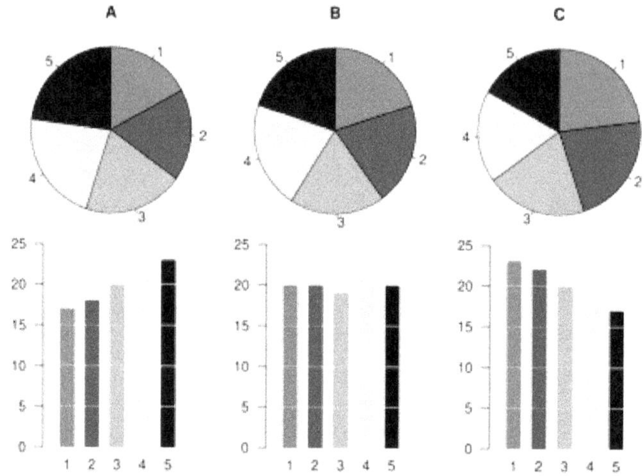

Fig 8: Three Sets of Percentages, Plotted as Both Piecharts and Barcharts. Comparing the Data on Barcharts is Generally Easier

Statisticians generally regard pie charts as a poor method of displaying information, and they are uncommon in scientific literature. One reason is that it is more difficult for comparisons to be made between the size of items in a chart when area is used instead of length and when different items are shown as different shapes.

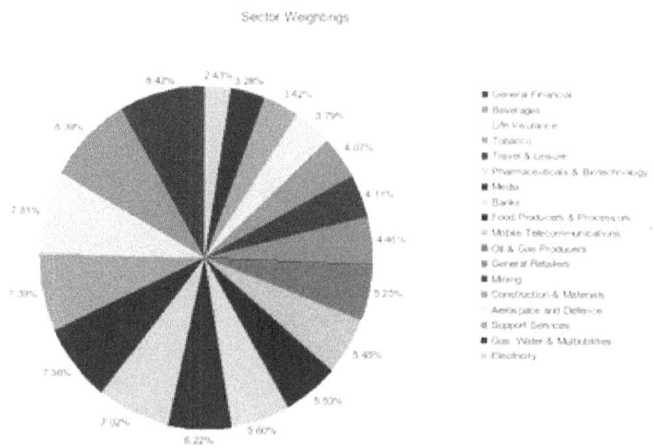

Fig 9: An Example of a Pie Chart with 18 Values, with Some Colors Repeated

Further, in research performed at AT&T Bell Laboratories, it was shown that comparison by angle was less accurate than comparison by length. Most subjects have difficulty ordering the slices in the pie chart by size; when an equivalent bar chart is used the comparison is much easier. Similarly, comparisons between data sets are easier using the bar chart. However, if the goal is to compare a given category (a slice of the pie) with the total (the whole pie) in a single chart and the multiple is close to 25 or 50 percent, then a pie chart can often be more effective than a bar graph. In a pie chart with many section, several values may be represented with the same or similar colors, making interpretation difficult. Several studies presented at the *European Visualization Conference* analyzed the relative accuracy of several pie chart formats, reaching the conclusion that pie charts and doughnut charts produce similar error levels when reading them, and square pie charts provide the most accurate reading.

Bar Chart

A bar chart or bar graph is a chart or graph that presents categorical data with rectangular bars with heights or lengths proportional to the values that they represent. The bars can be plotted vertically or horizontally. A vertical bar chart is sometimes called a column chart.

A bar graph shows comparisons among discrete categories. One axis of the chart shows the specific categories being compared, and the other axis represents a measured value. Some bar graphs present bars clustered in groups of more than one, showing the values of more than one measured variable.

History

Many sources consider William Playfair (1759-1824) to have invented the bar chart and the *Exports and Imports of Scotland to and from different parts for one Year from Christmas 1780 to Christmas 1781* graph from his *The Commercial and Political Atlas* to be the first bar chart in history. Diagrams of the velocity of a constantly accelerating object against time published in *The Latitude of Forms* (attributed to Jacobus de Sancto Martino or, perhaps, to Nicole Oresme) about 300 years before can be interpreted as "proto bar charts".

Usage

Bar graphs/charts provide a visual presentation of categorical data. Categorical data is a grouping of data into discrete groups, such as months of the year, age group, shoe sizes, and animals. These categories are usually qualitative. In a column (vertical) bar chart, categories appear along the horizontal axis and the height of the bar corresponds to the value of each category.

Bar charts have a discrete domain of categories, and are usually scaled so that all the data can fit on the chart. When there is no natural ordering of the categories being compared, bars on the chart may be arranged in any order. Bar charts arranged from highest to lowest incidence are called Pareto charts.

Grouped (Clustered) and Stacked

Bar graphs can also be used for more complex comparisons of data with grouped (or "clustered") bar charts, and stacked bar charts. In grouped (clustered) bar charts, for each categorical group there are two or more bars color-coded to represent a particular grouping. For example, a business owner with two stores might make a grouped bar chart with different colored bars to represent each store: the horizontal axis would show the months of the year and the vertical axis would show revenue. Alternatively, Stacked bar charts (also known as Composite bar charts) stack bars on top of each other so that the height of the resulting stack shows the combined result. Unlike a grouped bar chart where each factor is displayed next to another, each with their own bar, the stacked bar chart displays multiple data points stacked in a single row or column. This may, for instance, take the form of uniform height bars charting a time series with internal stacked colours indicating the percentage participation of a sub-type of data.

Another example would be a time series displaying total numbers, with internal colors indicating participation in the total by sub-types. Stacked bar charts are not suited to data sets having both positive and negative values. Grouped bar charts usually present the information in the same order in each grouping. Stacked bar charts present the information in the same sequence on each bar.

Variable-Width (Variwide)

Variable-width bar charts, sometimes abbreviated *variwide* (bar) charts, are bar charts having bars with non-uniform widths. Generally:

- Bars represent quantities with respective rectangles of areas A that are respective arithmetic products of related pairs of
  - vertical-axis quantities (A/X) and
  - horizontal-axis quantities (X).
- Arithmetically, the area of each bar (rectangle) is determined a product of sides' lengths:

$$(A/X) * X = Area\ A \text{ for each bar}$$

Roles of the vertical and horizontal axes may be reversed, depending on the desired application. Examples of variable-width bar charts are shown at Wikimedia Commons.

Advantages

- **Easy to Read and Interpret:** Bar charts are easy to read and interpret, even for people without a background in statistics or data visualization. The bars make it easy to compare values and see trends, making it a useful tool for communicating information to a wide range of audiences.
- **Can Handle Large Amounts of Data:** Bar charts can handle large amounts of data and still provide a clear representation of the information. The bars can be made narrow or wide to fit a large number

of categories or data points, and the use of color or patterns can make it easier to distinguish between them.

- **Customizable:** Bar charts can be customized to suit the needs of the user. For example, the color, width, and height of the bars can be adjusted to make the chart more visually appealing, and labels and annotations can be added to provide additional information.
- **Useful for Comparing Values:** Bar charts are particularly useful for comparing values between categories or data points. They allow for quick identification of differences and similarities, making it easy to draw conclusions and make decisions.

Limitations

- **Limited Use for Continuous Data:** Bar charts are not useful for displaying continuous data, such as temperature or time. For continuous data, a line chart or scatter plot may be more appropriate. Bar charts of continuous data with error bars are sometimes referred to as dynamite plots.
- **Limited Use for Small Sample Sizes:** Bar charts may not be useful for displaying small sample sizes, as the bars may not accurately represent the data. In such cases, a histogram or box plot may be more appropriate.
- **May be Misleading:** Bar charts can be misleading if the scale is not appropriate or if the data is presented in a way that is designed to mislead the viewer. For example, if the y-axis is truncated, the differences between the bars may appear larger than they actually are.
- **Limited Scope for Multivariate Data:** Bar charts can only display one or two variables at a time, making them less useful for displaying multivariate data. In such cases, a scatter plot or heat map may be more appropriate.

5.4 IMPORTANCE OF FREQUENCY DISTRIBUTION IN RESEARCH

5.4.1 Summarizing Data

Frequency distribution provides a way to organize raw data into a summary format, making it easier to understand large datasets. By showing how frequently each value or category occurs, researchers can quickly see the overall structure of the data.

5.4.2 Identifying Data Patterns

Frequency distributions help in identifying patterns and trends in the data. For example, they can show whether the data is symmetrically distributed or skewed, whether there are multiple peaks (modes), or if there are any outliers.

5.4.3 Facilitating Statistical Analysis

Frequency distributions are the foundation for many statistical analyses. Descriptive statistics, such as the mean and standard deviation, are based on the distribution of data. Inferential statistics, such as hypothesis testing and regression analysis, also rely on the underlying distribution of the data.

5.4.4 Informed Decision-Making

In research, frequency distributions allow for informed decision-making. Whether designing experiments, evaluating health interventions, or analyzing clinical data, understanding how variables are distributed helps researchers make evidence-based conclusions.

5.5 CONCLUSION

Frequency distribution is a fundamental concept in biostatistics and research methods, offering a structured way to summarize and analyze data. By organizing data into categories or intervals, researchers can better understand the distribution, patterns, and trends in their datasets. Whether presented in tabular form or graphically, frequency distributions serve as the foundation for many statistical analyses and are essential for drawing meaningful conclusions from data. In healthcare and medicine, frequency distributions play a critical role in guiding decision-making and improving patient outcomes through data-driven research.

CHAPTER 6

MEASURES OF CENTRAL TENDENCY AND LOCATION

Measures of central tendency and location are fundamental concepts in statistics and biostatistics that describe the central or typical value in a dataset. These measures are essential for summarizing data, as they provide a single value that represents the entire set of observations. Central tendency refers to the central point of a distribution, while location indicates the position of a specific value within the dataset. The most common measures of central tendency are the mean, median, and mode, whereas measures of location include percentiles, quartiles, and the range.

6.1 MEAN

A mean is a numeric quantity representing the "center" of a collection of numbers and is intermediate to the extreme values of the set of numbers. There are several kinds of means (or "measures of central tendency") in mathematics, especially in statistics. Each attempts to summarize or typify a given group of data, illustrating the magnitude and sign of the data set. Which of these measures is most illuminating depends on what is being measured, and on context and purpose.

The *arithmetic mean,* also known as "arithmetic average", is the sum of the values divided by the number of values. The arithmetic mean of a set of numbers x_1, x_2, ..., x_n is typically denoted using an overhead bar, $\bar{x}$. If the numbers are from observing a sample of a larger group, the arithmetic mean is termed the *sample mean* ($\bar{x}$) to distinguish it from the group mean (or expected value) of the underlying distribution, denoted μ or μ_x.

Outside probability and statistics, a wide range of other notions of mean are often used in geometry and mathematical analysis; examples are given below.

6.1.1 Pythagorean Means

Arithmetic Mean (AM)

The arithmetic mean (or simply *mean* or *average*) of a list of numbers, is the sum of all of the numbers divided by their count. Similarly, the mean of a sample $x_1, x_2, \ldots, x_n$ usually denoted by $\bar{x}$, is the sum of the sampled values divided by the number of items in the sample.

$$\bar{x} = \frac{1}{n}\left(\sum_{i=1}^{n} x_i\right) = \frac{x_1 + x_2 + \cdots + x_n}{n}$$

For example, the arithmetic mean of five values: 4, 36, 45, 50, 75 is:

$$\frac{4+36+45+50+75}{5} = \frac{210}{5} = 42.$$

Geometric mean (GM)

The geometric mean is an average that is useful for sets of positive numbers, that are interpreted according to their product (as is the case with rates of growth) and not their sum (as is the case with the arithmetic mean):

$$\bar{x} = \left(\prod_{i=1}^{n} x_i\right)^{\frac{1}{n}} = (x_1 x_2 \cdots x_n)^{\frac{1}{n}}$$

For example, the geometric mean of five values: 4, 36, 45, 50, 75 is:

$$(4 \times 36 \times 45 \times 50 \times 75)^{\frac{1}{5}} = \sqrt[5]{24\,300\,000} = 30.$$

Harmonic Mean (HM)

The harmonic mean is an average which is useful for sets of numbers which are defined in relation to some unit, as in the case of speed (i.e., distance per unit of time):

$$\bar{x} = n\left(\sum_{i=1}^{n} \frac{1}{x_i}\right)^{-1}$$

For example, the harmonic mean of the five values: 4, 36, 45, 50, 75 is

$$\frac{5}{\frac{1}{4} + \frac{1}{36} + \frac{1}{45} + \frac{1}{50} + \frac{1}{75}} = \frac{5}{\frac{1}{3}} = 15.$$

If we have five pumps that can empty a tank of a certain size in respectively 4, 36, 45, 50, and 75 minutes, then the harmonic mean of 15 15 tells us that these five different pumps working together will pump at the same rate as much as five pumps that can each empty the tank in 15 15 minutes.

Relationship between AM, GM, and HM

AM, GM, and HM satisfy these inequalities:

$$\text{AM} \geq \text{GM} \geq \text{HM}$$

Equality holds if all the elements of the given sample are equal.

6.1.2 Statistical Location

In descriptive statistics, the mean may be confused with the median, mode or mid-range, as any of these may incorrectly be called an "average" (more formally, a measure of central tendency). The mean of a set of observations is the arithmetic average of the values; however, for skewed distributions, the mean is not necessarily the same as the middle value (median), or the most likely value (mode). For example, mean income is typically skewed upwards by a small number of people with very large incomes, so that the majority have an income lower than the mean. By contrast, the median income is the level at which half the population is below and half is above. The mode income is the most likely income and favors the larger number of people with lower incomes. While the median and mode are often more intuitive measures for such skewed data, many skewed distributions are in fact best described by their mean, including the exponential and Poisson distributions.

6.1.3 Weighted Arithmetic Mean

The weighted arithmetic mean (or weighted average) is used if one wants to combine average values from different sized samples of the same population:

$$\bar{x} = \frac{\sum_{i=1}^{n} w_i \bar{x}_i}{\sum_{i=1}^{n} w_i}.$$

In other applications, they represent a measure for the reliability of the influence upon the mean by the respective values.

6.1.4 Truncated Mean

Sometimes, a set of numbers might contain outliers (i.e., data values which are much lower or much higher than the others). Often, outliers are erroneous data caused by artifacts. In this case, one can use a truncated mean. It involves discarding given parts of the data at the top or the bottom end, typically an equal amount at each end and then taking the arithmetic mean of the remaining data. The number of values removed is indicated as a percentage of the total number of values.

6.1.5 Interquartile Mean

The interquartile mean is a specific example of a truncated mean. It is simply the arithmetic mean after removing the lowest and the highest quarter of values.

$$\bar{x} = \frac{2}{n} \sum_{i=\frac{n}{4}+1}^{\frac{3}{4}n} x_i$$

Assuming the values have been ordered, so is simply a specific example of a weighted mean for a specific set of weights.

6.1.6 Mean of Angles and Cyclical Quantities

Angles, times of day, and other cyclical quantities require modular arithmetic to add and otherwise combine numbers. In all these situations, there will not be a unique mean. For example, the times an hour before and after midnight are equidistant to both midnight and noon. It is also possible that no mean exists. Consider a color wheel—there is no mean to the set of all colors. In these situations, you must decide which mean is most useful. You can do this by adjusting the values before averaging, or by using a specialized approach for the mean of circular quantities.

6.1.7 Fréchet Mean

The Fréchet mean gives a manner for determining the "center" of a mass distribution on a surface or, more generally, Riemannian manifold. Unlike many other means, the Fréchet mean is defined on a space whose elements cannot necessarily be added together or multiplied by scalars. It is sometimes also known as the *Karcher mean* (named after Hermann Karcher).

6.1.8 Triangular Sets

In geometry, there are thousands of different definitions for the center of a triangle that can all be interpreted as the mean of a triangular set of points in the plane.

6.1.9 Swanson's Rule

This is an approximation to the mean for a moderately skewed distribution. It is used in hydrocarbon exploration and is defined as:

$$m = 0.3P_{10} + 0.4P_{50} + 0.3P_{90}$$

6.2 ARITHMETIC MEAN

In mathematics and statistics, the arithmetic mean arithmetic average, or just the *mean* or *average* (when the context is clear) is the sum of a collection of numbers divided by the count of numbers in the collection. The collection is often a set of results from an experiment, an observational study, or a survey. The term "arithmetic mean" is preferred in some mathematics and statistics contexts because it helps distinguish it from other types of means, such as geometric and harmonic.

In addition to mathematics and statistics, the arithmetic mean is frequently used in economics, anthropology, history, and almost every academic field to some extent. For example, per capita income is the arithmetic average income of a nation's population.

While the arithmetic mean is often used to report central tendencies, it is not a robust statistic: it is greatly influenced by outliers (values much larger or smaller than most others). For skewed distributions, such as the distribution of income for which a few people's incomes are substantially higher than most people's, the arithmetic mean may not coincide with one's notion of "middle". In that case, robust statistics, such as the median, may provide a better description of central tendency.

The arithmetic mean of a set of observed data is equal to the sum of the numerical values of each observation, divided by the total number of observations. Symbolically, for a data set consisting of the values $x_1,\ldots,x_n$, the arithmetic mean is defined by the formula:

$$\bar{x} = \frac{1}{n}\left(\sum_{i=1}^{n} x_i\right) = \frac{x_1 + x_2 + \ldots + x_n}{n}$$

For example, if the monthly salaries of 10 10 employees are $\{2500, 2700, 2400, 2300, 2550, 2650, 2750, 2450, 2600, 2400\}$, then the arithmetic mean is:

6.3 MEDIAN

The median of a set of numbers is the value separating the higher half from the lower half of a data sample, a population, or a probability distribution. For a data set, it may be thought of as the "middle" value. The basic feature of the median in describing data compared to the mean (often simply described as the "average") is that it is not skewed by a small proportion of extremely large or small values, and therefore provides a better representation of the center. Median income, for example, may be a better way to describe the center of the income distribution because increases in the largest incomes alone have no effect on the median. For this reason, the median is of central importance in robust statistics.

6.3.1 Finite Set of Numbers

The median of a finite list of numbers is the "middle" number, when those numbers are listed in order from smallest to greatest.

If the data set has an odd number of observations, the middle one is selected (after arranging in ascending order). For example, the following list of seven numbers,

$$1, 3, 3, 6, 7, 8, 9$$

has the median of 6, which is the fourth value.

If the data set has an even number of observations, there is no distinct middle value and the median is usually defined to be the arithmetic mean of the two middle values. For example, this data set of 8 numbers

$$1, 2, 3, 4, 5, 6, 8, 9$$

has a median value of 4.5, that is (4 + 5) / 2 $(4+5)/2$. (In more technical terms, this interprets the median as the fully trimmed mid-range).

6.3.2 Definition and Notation

Formally, a median of a population is any value such that at least half of the population is less than or equal to the proposed median and at least half is greater than or equal to the proposed median. Medians may not be unique. If each set contains more than half the population, then some of the population is exactly equal to the unique median.

The median is well-defined for any ordered (one-dimensional) data and is independent of any distance metric. The median can thus be applied to school classes which are ranked but not numerical (e.g. working out a median grade when student test scores are graded from F to A), although the result might be halfway between classes if there is an even number of classes. (For odd number classes, one specific class is determined as the median.)

A geometric median, on the other hand, is defined in any number of dimensions. A related concept, in which the outcome is forced to correspond to a member of the sample, is the medoid.

There is no widely accepted standard notation for the median, but some authors represent the median of a variable x as med(x), as $\mu_{1/2}$, or as M. In any of these cases, the use of these or other symbols for the median needs to be explicitly defined when they are introduced.

The median is a special case of other ways of summarizing the typical values associated with a statistical distribution: it is the 2nd quartile, 5th decile, and 50th percentile.

6.3.3 Uses

The median can be used as a measure of location when one attaches reduced importance to extreme values, typically because a distribution is skewed, extreme values are not known, or outliers are untrustworthy, i.e., may be measurement or transcription errors. For example, consider the multiset

$$1, 2, 2, 2, 3, 14.$$

The median is 2 in this case, as is the mode, and it might be seen as a better indication of the center than the arithmetic mean of 4, which is larger than all but one of the values. However, the widely cited empirical relationship that the mean is shifted "further into the tail" of a distribution than the median is not generally true.

As a median is based on the middle data in a set, it is not necessary to know the value of extreme results in order to calculate it. For example, in a psychology test investigating the time needed to solve a problem, if a small number of people failed to solve the problem at all in the given time a median can still be calculated.

Because the median is simple to understand and easy to calculate, while also a robust approximation to the mean, the median is a popular summary statistic in descriptive statistics. In this context, there are several choices for a measure of variability: the range, the interquartile range, the mean absolute deviation, and the median absolute deviation.

For practical purposes, different measures of location and dispersion are often compared on the basis of how well the corresponding population values can be estimated from a sample of data. The median, estimated using the sample median, has good properties in this regard. While it is not usually optimal if a given population distribution is assumed, its properties are always reasonably good. For example, a comparison of the efficiency of candidate estimators shows that the sample mean is more statistically efficient when—and only when— data is uncontaminated by data from heavy-tailed distributions or from mixtures of distributions. Even then, the median has a 64% efficiency compared to the minimum-variance mean (for large normal samples), which is to say the variance of the median will be ~50% greater than the variance of the mean.

6.4 MODE

In statistics, the mode is the value that appears most often in a set of data values. If X is a discrete random variable, the mode is the value x at which the probability mass function takes its maximum value (i.e., $x = \text{argmax}_{x_i} P(X = x_i)$). In other words, it is the value that is most likely to be sampled.

Like the statistical mean and median, the mode is a way of expressing, in a (usually) single number, important information about a random variable or a population. The numerical value of the mode is the same as that of the mean and median in a normal distribution, and it may be very different in highly skewed distributions.

The mode is not necessarily unique in a given discrete distribution since the probability mass function may take the same maximum value at several points x_1, x_2, etc. The most extreme case occurs in uniform distributions, where all values occur equally frequently.

A mode of a continuous probability distribution is often considered to be any value x at which its probability density function has a locally maximum value. When the probability density function of a continuous distribution has multiple local maxima it is common to refer to all of the local maxima as modes of the distribution, so any peak is a mode. Such a continuous distribution is called multimodal (as opposed to unimodal).

In symmetric unimodal distributions, such as the normal distribution, the mean (if defined), median and mode all coincide. For samples, if it is known that they are drawn from a symmetric unimodal distribution, the sample mean can be used as an estimate of the population mode.

6.4.1 Mode of a Sample

The mode of a sample is the element that occurs most often in the collection. For example, the mode of the sample [1, 3, 6, 6, 6, 6, 7, 7, 12, 12, 17] is 6. Given the list of data [1, 1, 2, 4, 4] its mode is not unique. A dataset, in such a case, is said to be bimodal, while a set with more than two modes may be described as multimodal.

For a sample from a continuous distribution, such as [0.935..., 1.211..., 2.430..., 3.668..., 3.874...], the concept is unusable in its raw form, since no two values will be exactly the same, so each value will occur precisely once. In order to estimate the mode of the underlying distribution, the usual practice is to discretize the data by assigning frequency values to intervals of equal distance, as for making a histogram, effectively replacing the values by the midpoints of the intervals they are assigned to. The mode is then the value where the histogram reaches its peak. For small or middle-sized samples the outcome of this procedure is sensitive to the choice of interval width if chosen too narrow or too wide; typically one should have a sizable fraction of the data concentrated in a relatively small number of intervals (5 to 10), while the fraction of the data falling outside these intervals is also sizable. An alternate approach is kernel density estimation, which essentially blurs point samples to produce a continuous estimate of the probability density function which can provide an estimate of the mode.

The following MATLAB (or Octave) code example computes the mode of a sample:

```
X = sort(x);                    % x is a column vector dataset
indices  = find(diff([X; realmax]) > 0); % indices where repeated values change
[modeL,i] = max (diff([0; indices]));   % longest persistence length of repeated values
mode     = X(indices(i));
```

The algorithm requires as a first step to sort the sample in ascending order. It then computes the discrete derivative of the sorted list and finds the indices where this derivative is positive. Next it computes the discrete derivative of this set of indices, locating the maximum of this derivative of indices, and finally evaluates the sorted sample at the point where that maximum occurs, which corresponds to the last member of the stretch of repeated values.

6.4.2 Comparison of Mean, Median and Mode

Use

Unlike mean and median, the concept of mode also makes sense for "nominal data" (i.e., not consisting of numerical values in the case of mean, or even of ordered values in the case of median).

For example, taking a sample of Korean family names, one might find that "Kim" occurs more often than any other name. Then "Kim" would be the mode of the sample. In any voting system where a plurality determines victory, a single modal value determines the victor, while a multi-modal outcome would require some tie-breaking procedure to take place.

Unlike median, the concept of mode makes sense for any random variable assuming values from a vector space, including the real numbers (a one-dimensional vector space) and the integers (which can be considered embedded in the reals). For example, a distribution of points in the plane will typically have a mean and a mode, but the concept of median does not apply. The median makes sense when there is a linear order on the possible values. Generalizations of the concept of median to higher-dimensional spaces are the geometric median and the centerpoint.

Uniqueness and Definedness

For some probability distributions, the expected value may be infinite or undefined, but if defined, it is unique. The mean of a (finite) sample is always defined. The median is the value such that the fractions not exceeding it and not falling below it are each at least 1/2. It is not necessarily unique, but never infinite or totally undefined. For a data sample it is the "halfway" value when the list of values is ordered in increasing value, where usually for a list of even length the numerical average is taken of the two values closest to "halfway". Finally, as said before, the mode is not necessarily unique. Certain pathological distributions (for example, the Cantor distribution) have no defined mode at all. For a finite data sample, the mode is one (or more) of the values in the sample.

Properties

Assuming definedness, and for simplicity uniqueness, the following are some of the most interesting properties.

- All three measures have the following property: If the random variable (or each value from the sample) is subjected to the linear or affine transformation, which replaces X by $aX + b$, so are the mean, median and mode.
- Except for extremely small samples, the mode is insensitive to "outliers" (such as occasional, rare, false experimental readings). The median is also very robust in the presence of outliers, while the mean is rather sensitive.
- In continuous unimodal distributions the median often lies between the mean and the mode, about one third of the way going from mean to mode. In a formula, median ≈ (2 × mean + mode)/3. This rule, due to Karl Pearson, often applies to slightly non-symmetric distributions that resemble a normal distribution, but it is not always true and in general the three statistics can appear in any order.
- For unimodal distributions, the mode is within $\sqrt{3}$ standard deviations of the mean, and the root mean square deviation about the mode is between the standard deviation and twice the standard deviation.

6.5 PERCENTILE

In statistics, a *k*-th percentile, also known as percentile score or centile, is a score *below which* a given percentage *k* of scores in its frequency distribution falls ("exclusive" definition) or a score *at or below which* a given percentage falls ("inclusive" definition). Percentiles are expressed in the same unit of measurement as the input scores, *not* in percent; for example, if the scores refer to human weight, the corresponding percentiles will be expressed in kilograms or pounds. In the limit of an infinite sample size, the percentile approximates the *percentile function*, the inverse of the cumulative distribution function.

Percentiles are a type of quantiles, obtained adopting a subdivision into 100 groups. The 25th percentile is also known as the first *quartile* (Q_1), the 50th percentile as the *median* or second quartile (Q_2), and the 75th percentile as the third quartile (Q_3). For example, the 50th percentile (median) is the score *below* (or *at or below*, depending on the definition) which 50% of the scores in the distribution are found.

A related quantity is the *percentile rank* of a score, expressed in percent, which represents the fraction of scores in its distribution that are less than it, an exclusive definition. Percentile scores and percentile ranks are often used in the reporting of test scores from norm-referenced tests, but, as just noted, they are not the same. For percentile ranks, a score is given and a percentage is computed. Percentile ranks are exclusive: if the percentile rank for a specified score is 90%, then 90% of the scores were lower. In contrast, for percentiles a percentage is given and a corresponding score is determined, which can be either exclusive or inclusive. The score for a specified percentage (e.g., 90th) indicates a score below which (exclusive definition) or at or below which (inclusive definition) other scores in the distribution fall.

6.5.1 Definitions

There is no standard definition of percentile; however, all definitions yield similar results when the number of observations is very large and the probability distribution is continuous. In the limit, as the sample size approaches infinity, the $100p^{th}$ percentile ($0<p<1$) approximates the inverse of the cumulative distribution function (CDF) thus formed, evaluated at p, as p approximates the CDF. This can be seen as a consequence of the Glivenko–Cantelli theorem. Some methods for calculating the percentiles are given below.

6.5.2 The Normal Distribution and Percentiles

The methods given in the *calculation methods* section (below) are approximations for use in small-sample statistics. In general terms, for very large populations following a normal distribution, percentiles may often be represented by reference to a normal curve plot. The normal distribution is plotted along an axis scaled to standard deviations, or sigma (σ) units. Mathematically, the normal distribution extends to negative infinity on the left and positive infinity on the right. Note, however, that only a very small proportion of individuals in a population will fall outside the -3σ to +3σ range. For example, with human heights very few people are above the +3σ height level.

Percentiles represent the area under the normal curve, increasing from left to right. Each standard deviation represents a fixed percentile. Thus, rounding to two decimal places, -3σ is the 0.13th percentile, -2σ the 2.28th percentile, -1σ the 15.87th percentile, 0σ the 50th percentile (both the mean and median of the distribution), +1σ the 84.13th percentile, +2σ the 97.72nd percentile, and +3σ the 99.87th percentile. This is related to the 68–95–99.7 rule or the three-sigma rule. Note that in theory the 0th percentile falls at negative infinity and the 100th percentile at positive infinity, although in many practical applications, such as test results, natural lower and/or upper limits are enforced.

6.5.3 Applications

When ISPs bill "burstable" internet bandwidth, the 95th or 98th percentile usually cuts off the top 5% or 2% of bandwidth peaks in each month, and then bills at the nearest rate. In this way, infrequent peaks are ignored, and the customer is charged in a fairer way. The reason this statistic is so useful in measuring data throughput is that it gives a very accurate picture of the cost of the bandwidth. The 95th percentile says that 95% of the time, the usage is below this amount: so, the remaining 5% of the time, the usage is above that amount.

Physicians will often use infant and children's weight and height to assess their growth in comparison to national averages and percentiles which are found in growth charts.

The 85th percentile speed of traffic on a road is often used as a guideline in setting speed limits and assessing whether such a limit is too high or low.

In finance, value at risk is a standard measure to assess (in a model-dependent way) the quantity under which the value of the portfolio is not expected to sink within a given period of time and given a confidence value.

6.6 QUARTILE

In statistics, quartiles are a type of quantiles which divide the number of data points into four parts, or *quarters*, of more-or-less equal size. The data must be ordered from smallest to largest to compute quartiles; as such, quartiles are a form of order statistic. The three quartiles, resulting in four data divisions, are as follows:

- **The First Quartile (Q_1)** is defined as the 25th percentile where lowest 25% data is below this point. It is also known as the *lower* quartile.

- **The Second Quartile (Q_2)** is the median of a data set; thus 50% of the data lies below this point.
- **The Third Quartile (Q_3)** is the 75th percentile where lowest 75% data is below this point. It is known as the *upper* quartile, as 75% of the data lies below this point.

Along with the minimum and maximum of the data (which are also quartiles), the three quartiles described above provide a five-number summary of the data. This summary is important in statistics because it provides information about both the center and the spread of the data.

Knowing the lower and upper quartile provides information on how big the spread is and if the dataset is skewed toward one side. Since quartiles divide the number of data points evenly, the range is generally not the same between adjacent quartiles (i.e. usually $(Q_3 - Q_2) \neq (Q_2 - Q_1)$). Interquartile range (IQR) is defined as the difference between the 75th and 25th percentiles or $Q_3 - Q_1$. While the maximum and minimum also show the spread of the data, the upper and lower quartiles can provide more detailed information on the location of specific data points, the presence of outliers in the data, and the difference in spread between the middle 50% of the data and the outer data points.

6.6.1 Discrete Distributions

For discrete distributions, there is no universal agreement on selecting the quartile values.

Method 1

- Use the median to divide the ordered data set into two halves. The median becomes the second quartile.
  - If there are an odd number of data points in the original ordered data set, do not include the median (the central value in the ordered list) in either half.
  - If there are an even number of data points in the original ordered data set, split this data set exactly in half.
- The lower quartile value is the median of the lower half of the data. The upper quartile value is the median of the upper half of the data.

This rule is employed by the TI-83 calculator boxplot and "1-Var Stats" functions.

Method 2

- Use the median to divide the ordered data set into two halves. The median becomes the second quartile.
  - If there are an odd number of data points in the original ordered data set, include the median (the central value in the ordered list) in both halves.

   - If there are an even number of data points in the original ordered data set, split this data set exactly in half.
- The lower quartile value is the median of the lower half of the data. The upper quartile value is the median of the upper half of the data.

The values found by this method are also known as "Tukey's hinges".

Method 3

- Use the median to divide the ordered data set into two halves. The median becomes the second quartiles.
   - If there are odd numbers of data points, then go to the next step.
   - If there are even numbers of data points, then the Method 3 starts off the same as the Method 1 or the Method 2 above and you can choose to include or not include the median as a new datapoint. If you choose to include the median as the new datapoint, then proceed to the step 2 or 3 below because you now have an odd number of datapoints. If you do not choose the median as the new data point, then continue the Method 1 or 2 where you have started.
- If there are $(4n+1)$ data points, then the lower quartile is 25% of the nth data value plus 75% of the $(n+1)$th data value; the upper quartile is 75% of the $(3n+1)$th data point plus 25% of the $(3n+2)$th data point.
- If there are $(4n+3)$ data points, then the lower quartile is 75% of the $(n+1)$th data value plus 25% of the $(n+2)$th data value; the upper quartile is 25% of the $(3n+2)$th data point plus 75% of the $(3n+3)$th data point.

6.7 RANGE

In descriptive statistics, the range of a set of data is size of the narrowest interval which contains all the data. It is calculated as the difference between the largest and smallest values (also known as the sample maximum and minimum). It is expressed in the same units as the data.

The range provides an indication of statistical dispersion. Since it only depends on two of the observations, it is most useful in representing the dispersion of small data sets.

6.8 CONCLUSION

Measures of central tendency and location are vital tools for summarizing and interpreting data in research and biostatistics.

The mean, median, and mode each provide a different perspective on the central point of a dataset, while percentiles, quartiles, and the range offer insights into the distribution and location of values.

By choosing the appropriate measure based on the characteristics of the data, researchers can gain meaningful insights and make informed decisions. Understanding these measures is essential for analyzing data in a wide range of fields, from medicine and public health to economics and social sciences.

CHAPTER 7

MEASURES OF DISPERSION

In the field of statistics, measures of central tendency such as the mean, median, and mode are important for identifying the central value or typical data point in a dataset. However, these measures do not provide any insight into the variability or spread of the data. This is where measures of dispersion, also known as measures of variability, come into play. They help quantify the extent to which data points in a dataset are spread out or clustered around the central value. Understanding the variability in a dataset is crucial for making informed decisions, particularly in areas like health research, business analytics, and social sciences.

There are several different measures of dispersion, including range, variance, standard deviation, interquartile range, and mean absolute deviation. Each of these methods offers a unique way to understand data variability. In this discussion, we will explore these key measures of dispersion, their importance, how they are calculated, and their applications in research and decision-making.

7.1 IMPORTANCE OF MEASURES OF DISPERSION

Measures of dispersion are critical for understanding the reliability and consistency of data. Two datasets may have the same mean but very different spreads of values, which can lead to different interpretations and conclusions. For instance, in medical research, understanding the variability in patients' responses to a drug treatment can help identify outliers or extreme cases that require special attention.

7.1.1 Applications in Different Fields

- **Healthcare:** Dispersion helps in analyzing the spread of clinical measurements, such as blood pressure or cholesterol levels, across different patient populations. It can guide treatment plans by identifying which patients fall outside the normal range.
- **Business:** In market analysis, measures of dispersion are used to assess the volatility of stock prices or the variation in customer preferences, helping businesses adapt to market conditions.
- **Social Sciences:** Variability in income distribution, educational attainment, or social behaviors can be assessed using measures of dispersion, offering deeper insights into societal inequalities or trends.

7.2 RANGE

The range is the simplest measure of dispersion. It is calculated by subtracting the smallest value in the dataset from the largest value. Mathematically, it can be expressed as:

Range = Maximum Value - Minimum Value

While the range provides a quick sense of the spread in the data, it is highly sensitive to extreme values or outliers. A dataset with one very high or very low value can have a large range, even if most of the data points are clustered together.

7.2.1 Advantages of the Range

- **Simple to Calculate:** The range is easy to compute and understand, making it a useful starting point for assessing data dispersion.
- **Highlights Extremes:** The range gives immediate information about the highest and lowest values in a dataset, which can be useful for identifying potential outliers.

7.2.2 Disadvantages of the Range

- **Sensitive to Outliers:** Since the range only considers the two extreme values, it can be misleading if the dataset contains outliers.
- **Ignores Overall Distribution:** The range does not take into account the distribution of values between the extremes, making it a less reliable measure of dispersion for complex datasets.

7.3 VARIANCE

Variance is a more comprehensive measure of dispersion that considers all the values in the dataset. It represents the average of the squared differences between each data point and the mean. The formula for variance is:

$$Variance(\sigma^2) = \frac{\sum(X-\mu)^2}{n}$$

Where:

- X represents each individual data point,
- μ is the mean of the dataset, and
- n is the number of observations.

By squaring the differences, variance ensures that positive and negative deviations from the mean do not cancel each other out.

A higher variance indicates a greater spread of data around the mean, while a lower variance suggests that the data points are closely clustered around the mean.

7.3.1 Advantages of Variance

- **Comprehensive Measure:** Variance takes into account every data point in the dataset, providing a thorough representation of variability.
- **Mathematically Important:** Variance is used in many advanced statistical techniques, including hypothesis testing and regression analysis.

7.3.2 Disadvantages of Variance

- **Difficult to Interpret:** Because variance is expressed in squared units, it can be hard to interpret directly, especially in practical situations where the original units (e.g., meters, seconds) matter.
- **Sensitive to Outliers:** Like the range, variance can be affected by extreme values, which can distort the true measure of variability in the data.

7.4 STANDARD DEVIATION

The standard deviation is the square root of the variance and is one of the most commonly used measures of dispersion. Unlike variance, the standard deviation is expressed in the same units as the original data, making it more interpretable. The formula for standard deviation is:

$$\text{Standard Derivation } (\sigma) = \sqrt{\frac{\sum (X-\mu)^2}{n}}$$

Since standard deviation is based on variance, it shares the same strengths, including its ability to account for every data point in the dataset. It is widely used in research because it provides a clear and easy-to-interpret measure of data spread.

7.4.1 Advantages of Standard Deviation

- **Easy to Interpret:** Since it is expressed in the same units as the data, standard deviation is more interpretable than variance.
- **Widely Used:** Standard deviation is a standard measure in many statistical techniques, including normal distribution analysis and confidence intervals.

7.4.2 Disadvantages of Standard Deviation

- **Affected by Outliers:** Like variance, standard deviation is sensitive to outliers, which can inflate the measure of dispersion.
- **Not Ideal for Skewed Distributions:** Standard deviation assumes a normal distribution of data, so it may not be the best measure for skewed datasets.

7.5 INTERQUARTILE RANGE (IQR)

The interquartile range (IQR) is a measure of dispersion that focuses on the middle 50% of the data. It is the difference between the third quartile (Q3) and the first quartile (Q1):

$$IQR = Q3 - Q1$$

The IQR is particularly useful for skewed data or datasets with outliers because it ignores the extreme values and focuses on the central portion of the data. This makes it a robust measure of variability, especially when the range or standard deviation is affected by outliers.

7.5.1 Advantages of IQR

- **Resistant to Outliers:** By focusing on the middle 50% of the data, the IQR is not affected by outliers or extreme values.
- **Ideal for Skewed Data:** The IQR is a good measure of dispersion for skewed distributions, where other measures like the standard deviation may not provide accurate insights.

7.5.2 Disadvantages of IQR

- **Ignores Extremes:** While the IQR is resistant to outliers, it also ignores valuable information from the extremes of the dataset, which could be important in certain analyses.
- **Less Common in Some Fields:** The IQR is less widely used in comparison to variance and standard deviation, especially in parametric statistical methods.

7.6 MEAN ABSOLUTE DEVIATION (MAD)

The mean absolute deviation (MAD) is another measure of dispersion that calculates the average of the absolute differences between each data point and the mean. Unlike variance, MAD does not square the differences, making it a more intuitive measure for some applications. The formula for MAD is:

$$MAD = \frac{\sum |X - \mu|}{n}$$

7.6.1 Advantages of MAD

- **Easy to Understand:** MAD is more intuitive than variance because it uses absolute differences rather than squared differences.
- **Less Sensitive to Outliers:** While MAD is still affected by outliers, it is less sensitive than variance or standard deviation because it does not square the differences.

7.6.2 Disadvantages of MAD

- **Less Precise:** MAD may be less precise than variance or standard deviation for datasets with complex distributions.
- **Less Commonly Used:** MAD is less commonly used in statistical analysis, particularly in fields that rely on more advanced statistical techniques.

7.7 USES

The median absolute deviation is a measure of statistical dispersion. Moreover, the MAD is a robust statistic, being more resilient to outliers in a data set than the standard deviation.

In the standard deviation, the distances from the mean are squared, so large deviations are weighted more heavily, and thus outliers can heavily influence it. In the MAD, the deviations of a small number of outliers are irrelevant. Because the MAD is a more robust estimator of scale than the sample variance or standard deviation, it works better with distributions without a mean or variance, such as the Cauchy distribution.

7.8 MAD USING GEOMETRIC MEDIAN

Analogously to how the median generalizes to the geometric median (gm) in multivariate data, MAD can be generalized to MADGM (median of distances to gm) in n dimensions. This is done by replacing the absolute differences in one dimension by euclidean distances of the data points to the geometric median in n dimensions. This gives the identical result as the univariate MAD in 1 dimension and generalizes to any number of dimensions. MADGM needs the geometric median to be found, which is done by an iterative process.

7.9 THE POPULATION MAD

The population MAD is defined analogously to the sample MAD, but is based on the complete distribution rather than on a sample. For a symmetric distribution with zero mean, the population MAD is the 75th percentile of the distribution.

Unlike the variance, which may be infinite or undefined, the population MAD is always a finite number. For example, the standard Cauchy distribution has undefined variance, but its MAD is 1. The earliest known mention of the concept of the MAD occurred in 1816, in a paper by Carl Friedrich Gauss on the determination of the accuracy of numerical observations.

7.10 CONCLUSION

Measures of dispersion are essential for understanding the variability and spread of data, complementing measures of central tendency. From the simplicity of the range to the more complex standard deviation and IQR, each measure has its strengths and limitations, making them suitable for different types of data and research questions. Selecting the appropriate measure depends on the dataset's characteristics, such as the presence of outliers or whether the data is normally distributed.

By understanding the various measures of dispersion, researchers and analysts can better interpret data and make informed decisions in fields ranging from healthcare to economics.

CHAPTER 8
PROBABILITY

Probability is the branch of mathematics concerning events and numerical descriptions of how likely they are to occur. The probability of an event is a number between 0 and 1; the larger the probability, the more likely an event is to occur. A simple example is the tossing of a fair (unbiased) coin. Since the coin is fair, the two outcomes ("heads" and "tails") are both equally probable; the probability of "heads" equals the probability of "tails"; and since no other outcomes are possible, the probability of either "heads" or "tails" is 1/2 (which could also be written as 0.5 or 50%).

These concepts have been given an axiomatic mathematical formalization in *probability theory*, which is used widely in areas of study such as statistics, mathematics, science, finance, gambling, artificial intelligence, machine learning, computer science, game theory, and philosophy to, for example, draw inferences about the expected frequency of events. Probability theory is also used to describe the underlying mechanics and regularities of complex systems.

8.1 HISTORY OF PROBABILITY

Probability has a dual aspect: on the one hand the likelihood of hypotheses given the evidence for them, and on the other hand the behavior of stochastic processes such as the throwing of dice or coins. The study of the former is historically older in, for example, the law of evidence, while the mathematical treatment of dice began with the work of Cardano, Pascal, Fermat and Christiaan Huygens between the 16th and 17th century. Probability deals with random experiments with a known distribution, Statistics deals with inference from the data about the unknown distribution.

8.1.1 Origins

Ancient and medieval law of evidence developed a grading of degrees of proof, credibility, presumptions and half-proof to deal with the uncertainties of evidence in court.

In Renaissance times, betting was discussed in terms of odds such as "ten to one" and maritime insurance premiums were estimated based on intuitive risks, but there was no theory on how to calculate such odds or premiums. The mathematical methods of probability arose in the investigations first of Gerolamo Cardano in the 1560s (not published until 100 years later), and then in the correspondence Pierre de Fermat and Blaise Pascal (1654) on such questions as the fair division of the stake in an interrupted game of chance. Christiaan Huygens (1657) gave a comprehensive treatment of the subject.

In ancient times there were games played using astragali, or talus bone. The Pottery of ancient Greece provides evidence to show that the astragali were tossed into a circle drawn on the floor, much like playing marbles. In Egypt, excavators of tombs found a game they called "Hounds and Jackals", which closely resembles the modern game "Snakes and Ladders".

According to Pausanias, Palamedes invented dice during the Trojan wars, although their true origin is uncertain. The first dice game mentioned in literature of the Christian era was called Hazard. Played with two or three dice, it was probably brought to Europe by the knights returning from the Crusades. Dante Alighieri (1265–1321) mentions this game.

A commenter of Dante puts further thought into this game: the thought was that with three dice, the lowest number you can get is three, an ace for every die. Achieving a four can be done with three dice by having a two on one die and aces on the other two dice.

Cardano also thought about the sum of three dice. At face value there are the same number of combinations that sum to 9 as those that sum to 10. For a 9:(621) (531) (522) (441) (432) (333) and for 10: (631) (622) (541) (532) (442) (433). However, there are more ways of obtaining some of these combinations than others. For example, if we consider the order of results there are six ways to obtain (621): (1,2,6), (1,6,2), (2,1,6), (2,6,1), (6,1,2), (6,2,1), but there is only one way to obtain (333), where the first, second and third dice all roll 3.

There are a total of 27 permutations that sum to 10 but only 25 that sum to 9. From this, Cardano found that the probability of throwing a 9 is less than that of throwing a 10. He also demonstrated the efficacy of defining odds as the ratio of favourable to unfavourable outcomes (which implies that the probability of an event is given by the ratio of favourable outcomes to the total number of possible outcomes).

In addition, Galileo wrote about die-throwing sometime between 1613 and 1623. Unknowingly considering what is essentially the same problem as Cardano's, Galileo had said that certain numbers have the ability to be thrown because there are more ways to create that number.

8.1.2 Eighteenth Century

Jacob Bernoulli's *Ars Conjectandi* (posthumous, 1713) and Abraham De Moivre's *The Doctrine of Chances* (1718) put probability on a sound mathematical footing, showing how to calculate a wide range of complex probabilities. Bernoulli proved a version of the fundamental law of large numbers, which states that in a large number of trials, the average of the outcomes is likely to be very close to the expected value - for example, in 1000 throws of a fair coin, it is likely that there are close to 500 heads (and the larger the number of throws, the closer to half-and-half the proportion is likely to be).

8.1.3 Nineteenth Century

The power of probabilistic methods in dealing with uncertainty was shown by Gauss's determination of the orbit of Ceres from a few observations. The theory of errors used the method of least squares to correct error-prone observations, especially in astronomy, based on the assumption of a normal distribution of errors to determine the most likely true value.

In 1812, Laplace issued his *Théorie analytique des probabilités* in which he consolidated and laid down many fundamental results in probability and statistics such as the moment-generating function, method of least squares, inductive probability, and hypothesis testing.

Towards the end of the nineteenth century, a major success of explanation in terms of probabilities was the Statistical mechanics of Ludwig Boltzmann and J. Willard Gibbs which explained properties of gases such as temperature in terms of the random motions of large numbers of particles.

The field of the history of probability itself was established by Isaac Todhunter's monumental *A History of the Mathematical Theory of Probability from the Time of Pascal to that of Laplace* (1865).

8.1.4 Twentieth Century

Probability and statistics became closely connected through the work on hypothesis testing of R. A. Fisher and Jerzy Neyman, which is now widely applied in biological and psychological experiments and in clinical trials of drugs, as well as in economics and elsewhere. A hypothesis, for example that a drug is usually effective, gives rise to a probability distribution that would be observed if the hypothesis is true. If observations approximately agree with the hypothesis, it is confirmed, if not, the hypothesis is rejected.

The theory of stochastic processes broadened into such areas as Markov processes and Brownian motion, the random movement of tiny particles suspended in a fluid. That provided a model for the study of random fluctuations in stock markets, leading to the use of sophisticated probability models in mathematical finance, including such successes as the widely used Black–Scholes formula for the valuation of options.

The twentieth century also saw long-running disputes on the interpretations of probability. In the mid-century frequentism was dominant, holding that probability means long-run relative frequency in a large number of trials. At the end of the century there was some revival of the Bayesian view, according to which the fundamental notion of probability is how well a proposition is supported by the evidence for it.

The mathematical treatment of probabilities, especially when there are infinitely many possible outcomes, was facilitated by Kolmogorov's axioms (1933).

8.2 INTERPRETATIONS

The word probability has been used in a variety of ways since it was first applied to the mathematical study of games of chance. Does probability measure the real, physical, tendency of something to occur, or is it a measure of how strongly one believes it will occur, or does it draw on both these elements? In answering such questions, mathematicians interpret the probability values of probability theory.

There are two broad categories of probability interpretations which can be called "physical" and "evidential" probabilities. Physical probabilities, which are also called objective or frequency probabilities, are associated with random physical systems such as roulette wheels, rolling dice and radioactive atoms. In such systems, a given type of event (such as a die yielding a six) tends to occur at a persistent rate, or "relative frequency", in a long run of trials. Physical probabilities either explain, or are invoked to explain, these stable frequencies. The two main kinds of theory of physical probability are frequentist accounts (such as those of Venn, Reichenbach and von Mises) and propensity accounts (such as those of Popper, Miller, Giere and Fetzer).

Evidential probability, also called Bayesian probability, can be assigned to any statement whatsoever, even when no random process is involved, as a way to represent its subjective plausibility, or the degree to which the statement is supported by the available evidence. On most accounts, evidential probabilities are considered to be degrees of belief, defined in terms of dispositions to gamble at certain odds. The four main evidential interpretations are the classical (e.g. Laplace's) interpretation, the subjective interpretation (de Finetti and Savage), the epistemic or inductive interpretation (Ramsey, Cox) and the logical interpretation (Keynes and Carnap). There are also evidential interpretations of probability covering groups, which are often labelled as 'intersubjective' (proposed by Gillies and Rowbottom).

Some interpretations of probability are associated with approaches to statistical inference, including theories of estimation and hypothesis testing. The physical interpretation, for example, is taken by followers of "frequentist" statistical methods, such as Ronald Fisher, Jerzy Neyman and Egon Pearson. Statisticians of the opposing Bayesian school typically accept the frequency interpretation when it makes sense (although not as a definition), but there is less agreement regarding physical probabilities. Bayesians consider the calculation of evidential probabilities to be both valid and necessary in statistics.

The terminology of this topic is rather confusing, in part because probabilities are studied within a variety of academic fields. The word "frequentist" is especially tricky. To philosophers it refers to a particular theory of physical probability, one that has more or less been abandoned. To scientists, on the other hand, "frequentist probability" is just another name for physical (or objective) probability. Those who promote Bayesian inference view "frequentist statistics" as an approach to statistical inference that is based on the frequency interpretation of probability, usually relying on the law of large numbers and characterized by what is called 'Null Hypothesis Significance Testing' (NHST). Also the word "objective", as applied to probability, sometimes means exactly what "physical" means here, but is also used of evidential probabilities that are fixed by rational constraints, such as logical and epistemic probabilities.

- It is unanimously agreed that statistics depends somehow on probability. But, as to what probability is and how it is connected with statistics, there has seldom been such complete disagreement and breakdown of communication since the Tower of Babel. Doubtless, much of the disagreement is merely terminological and would disappear under sufficiently sharp analysis.—*Savage, 1954, p. 2*

8.2.1 Philosophy

The philosophy of probability presents problems chiefly in matters of epistemology and the uneasy interface between mathematical concepts and ordinary language as it is used by non-mathematicians. Probability theory is an established field of study in mathematics. It has its origins in correspondence discussing the mathematics of games of chance between Blaise Pascal and Pierre de Fermat in the seventeenth century, and was formalized and rendered axiomatic as a distinct branch of mathematics by Andrey Kolmogorov in the twentieth century. In axiomatic form, mathematical statements about probability theory carry the same sort of epistemological confidence within the philosophy of mathematics as are shared by other mathematical statements.

The mathematical analysis originated in observations of the behaviour of game equipment such as playing cards and dice, which are designed specifically to introduce random and equalized elements; in mathematical terms, they are subjects of indifference. This is not the only way probabilistic statements are used in ordinary human language: when people say that "*it will probably rain*", they typically do not mean that the outcome of rain versus not-rain is a random factor that the odds currently favor; instead, such statements are perhaps better understood as qualifying their expectation of rain with a degree of confidence. Likewise, when it is written that "the most probable explanation" of the name of Ludlow, Massachusetts "is that it was named after Roger Ludlow", what is meant here is not that Roger Ludlow is favored by a random factor, but rather that this is the most plausible explanation of the evidence, which admits other, less likely explanations.

Thomas Bayes attempted to provide a logic that could handle varying degrees of confidence; as such, Bayesian probability is an attempt to recast the representation of probabilistic statements as an expression of the degree of confidence by which the beliefs they express are held.

Though probability initially had somewhat mundane motivations, its modern influence and use is widespread ranging from evidence-based medicine, through six sigma, all the way to the probabilistically checkable proof and the string theory landscape.

Table 1: A Summary of Some Interpretations of Probability

	Classical	Frequentist	Subjective	Propensity
Main hypothesis	Principle of indifference	Frequency of occurrence	Degree of belief	Degree of causal connection
Conceptual basis	Hypothetical symmetry	Past data and reference class	Knowledge and intuition	Present state of system
Conceptual approach	Conjectural	Empirical	Subjective	Metaphysical
Single case possible	Yes	No	Yes	Yes
Precise	Yes	No	No	Yes
Problems	Ambiguity in principle of indifference	Circular definition	Reference class problem	Disputed concept

8.2.2 Classical Definition

The first attempt at mathematical rigour in the field of probability, championed by Pierre-Simon Laplace, is now known as the classical definition. Developed from studies of games of chance (such as rolling dice) it states that probability is shared equally between all the possible outcomes, provided these outcomes can be deemed equally likely.

- The theory of chance consists in reducing all the events of the same kind to a certain number of cases equally possible, that is to say, to such as we may be equally undecided about in regard to their existence, and in determining the number of cases favorable to the event whose probability is sought. The ratio of this number to that of all the cases possible is the measure of this probability, which is thus simply a fraction whose numerator is the number of favorable cases and whose denominator is the number of all the cases possible.—*Pierre-Simon Laplace, A Philosophical Essay on Probabilities*

This can be represented mathematically as follows: If a random experiment can result in N mutually exclusive and equally likely outcomes and if N_A of these outcomes result in the occurrence of the event A, the probability of A is defined by

$$P(A) = \frac{N_A}{N}.$$

There are two clear limitations to the classical definition. Firstly, it is applicable only to situations in which there is only a 'finite' number of possible outcomes. But some important random experiments, such as tossing a coin until it shows heads, give rise to an infinite set of outcomes. And secondly, it requires an a priori determination that all possible outcomes are equally likely without falling in a trap of circular reasoning by relying on the notion of probability. (In using the terminology "we may be equally undecided", Laplace assumed, by what has been called the "principle of insufficient reason", that all possible outcomes are equally likely if there is no known reason to assume otherwise, for which there is no obvious justification.)

8.2.3 Frequentism

Frequentists posit that the probability of an event is its relative frequency over time, (3.4) i.e., its relative frequency of occurrence after repeating a process a large number of times under similar conditions. This is also known as aleatory probability.

The events are assumed to be governed by some random physical phenomena, which are either phenomena that are predictable, in principle, with sufficient information; or phenomena which are essentially unpredictable. Examples of the first kind include tossing dice or spinning a roulette wheel; an example of the second kind is radioactive decay. In the case of tossing a fair coin, frequentists say that the probability of getting a heads is 1/2, not because there are two equally likely outcomes but because repeated series of large numbers of trials demonstrate that the empirical frequency converges to the limit 1/2 as the number of trials goes to infinity.

If we denote by n a n_a the number of occurrences of an event $\mathcal{A}$ in n trials, then if lim n $\lim_{n\to+\infty} \frac{n_a}{n} = p$ we say that $P(\mathcal{A}) = p$.

The frequentist view has its own problems. It is of course impossible to actually perform an infinity of repetitions of a random experiment to determine the probability of an event. But if only a finite number of repetitions of the process are performed, different relative frequencies will appear in different series of trials. If these relative frequencies are to define the probability, the probability will be slightly different every time it is measured. But the real probability should be the same every time. If we acknowledge the fact that we only can measure a probability with some error of measurement attached, we still get into problems as the error of measurement can only be expressed as a probability, the very concept we are trying to define.

8.2.4 Subjectivism

Subjectivists, also known as Bayesians or followers of epistemic probability, give the notion of probability a subjective status by regarding it as a measure of the 'degree of belief' of the individual assessing the uncertainty of a particular situation. Epistemic or subjective probability is sometimes called credence, as opposed to the term chance for a propensity probability. Some examples of epistemic probability are to assign a probability to the proposition that a proposed law of physics is true or to determine how probable it is that a suspect committed a crime, based on the evidence presented. The use of Bayesian probability raises the philosophical debate as to whether it can contribute valid justifications of belief. Bayesians point to the work of Ramsey and de Finetti as proving that subjective beliefs must follow the laws of probability if they are to be coherent. Evidence casts doubt that humans will have coherent beliefs. The use of Bayesian probability involves specifying a prior probability. This may be obtained from consideration of whether the required prior probability is greater or lesser than a reference probability associated with an urn model or a thought experiment. The issue is that for a given problem, multiple thought experiments could apply, and choosing one is a matter of judgement: different people may assign different prior probabilities, known as the reference class problem. The "sunrise problem" provides an example.

8.2.5 Propensity

Propensity theorists think of probability as a physical propensity, or disposition, or tendency of a given type of physical situation to yield an outcome of a certain kind or to yield a long run relative frequency of such an outcome. This kind of objective probability is sometimes called 'chance'.

Propensities, or chances, are not relative frequencies, but purported causes of the observed stable relative frequencies. Propensities are invoked to explain why repeating a certain kind of experiment will generate given outcome types at persistent rates, which are known as propensities or chances. Frequentists are unable to take this approach, since relative frequencies do not exist for single tosses of a coin, but only for large ensembles or collectives. In contrast, a propensitist is able to use the law of large numbers to explain the behaviour of long-run frequencies. This law, which is a consequence of the axioms of probability, says that if (for example) a coin is tossed repeatedly many times, in such a way that its probability of landing heads is the same on each toss, and the outcomes are probabilistically independent, then the relative frequency of heads will be close to the probability of heads on each single toss. This law allows that stable long-run frequencies are a manifestation of invariant *single-case* probabilities. In addition to explaining the emergence of stable relative frequencies, the idea of propensity is motivated by the desire to make sense of single-case probability attributions in quantum mechanics, such as the probability of decay of a particular atom at a particular time.

The main challenge facing propensity theories is to say exactly what propensity means. (And then, of course, to show that propensity thus defined has the required properties.) At present, unfortunately, none of the well-recognised accounts of propensity comes close to meeting this challenge.

A propensity theory of probability was given by Charles Sanders Peirce. A later propensity theory was proposed by philosopher Karl Popper, who had only slight acquaintance with the writings of C. S. Peirce, however. Popper noted that the outcome of a physical experiment is produced by a certain set of "generating conditions". When we repeat an experiment, as the saying goes, we really perform another experiment with a (more or less) similar set of generating conditions. To say that a set of generating conditions has propensity p of producing the outcome E means that those exact conditions, if repeated indefinitely, would produce an outcome sequence in which E occurred with limiting relative frequency p. For Popper then, a deterministic experiment would have propensity 0 or 1 for each outcome, since those generating conditions would have same outcome on each trial. In other words, non-trivial propensities (those that differ from 0 and 1) only exist for genuinely nondeterministic experiments.

A number of other philosophers, including David Miller and Donald A. Gillies, have proposed propensity theories somewhat similar to Popper's.

Other propensity theorists (e.g. Ronald Giere) do not explicitly define propensities at all, but rather see propensity as defined by the theoretical role it plays in science. They argued, for example, that physical magnitudes such as electrical charge cannot be explicitly defined either, in terms of more basic things, but only in terms of what they do (such as attracting and repelling other electrical charges). In a similar way, propensity is whatever fills the various roles that physical probability plays in science.

What roles does physical probability play in science? What are its properties? One central property of chance is that, when known, it constrains rational belief to take the same numerical value. David Lewis called this the *Principal Principle*, (3.3 & 3.5) a term that philosophers have mostly adopted. For example, suppose you are certain that a particular biased coin has propensity 0.32 to land heads every time it is tossed. What is then the correct price for a gamble that pays $1 if the coin lands heads, and nothing otherwise? According to the Principal Principle, the fair price is 32 cents.

8.2.6 Logical, Epistemic, and Inductive Probability

It is widely recognized that the term "probability" is sometimes used in contexts where it has nothing to do with physical randomness. Consider, for example, the claim that the extinction of the dinosaurs was probably caused by a large meteorite hitting the earth. Statements such as "Hypothesis H is probably true" have been interpreted to mean that the (presently available) empirical evidence (E, say) supports H to a high degree. This degree of support of H by E has been called the logical, or epistemic, or inductive probability of H given E.

The differences between these interpretations are rather small, and may seem inconsequential. One of the main points of disagreement lies in the relation between probability and belief. Logical probabilities are conceived (for example in Keynes' Treatise on Probability) to be objective, logical relations between propositions (or sentences), and hence not to depend in any way upon belief. They are degrees of (partial) entailment, or degrees of logical consequence, not degrees of belief. (They do, nevertheless, dictate proper degrees of belief, as is discussed below.) Frank P. Ramsey, on the other hand, was skeptical about the existence of such objective logical relations and argued that (evidential) probability is "the logic of partial belief". (p 157) In other words, Ramsey held that epistemic probabilities simply *are* degrees of rational belief, rather than being logical relations that merely *constrain* degrees of rational belief.

Another point of disagreement concerns the *uniqueness* of evidential probability, relative to a given state of knowledge. Rudolf Carnap held, for example, that logical principles always determine a unique logical probability for any statement, relative to any body of evidence. Ramsey, by contrast, thought that while degrees of belief are subject to some rational constraints (such as, but not limited to, the axioms of probability) these constraints usually do not determine a unique value. Rational people, in other words, may differ somewhat in their degrees of belief, even if they all have the same information.

8.2.7 Prediction

An alternative account of probability emphasizes the role of *prediction* – predicting future observations on the basis of past observations, not on unobservable parameters. In its modern form, it is mainly in the Bayesian vein. This was the main function of probability before the 20th century, but fell out of favor compared to the parametric approach, which modeled phenomena as a physical system that was observed with error, such as in celestial mechanics.

The modern predictive approach was pioneered by Bruno de Finetti, with the central idea of exchangeability - that future observations should behave like past observations. This view came to the attention of the Anglophone world with the 1974 translation of de Finetti's book, and has since been propounded by such statisticians as Seymour Geisser.

8.2.8 Axiomatic Probability

The mathematics of probability can be developed on an entirely axiomatic basis that is independent of any interpretation.

8.3 PROBABILITY THEORY

Probability theory or probability calculus is the branch of mathematics concerned with probability. Although there are several different probability interpretations, probability theory treats the concept in a rigorous mathematical manner by expressing it through a set of axioms. Typically these axioms formalise probability in terms of a probability space, which assigns a measure taking values between 0 and 1, termed the probability measure, to a set of outcomes called the sample space. Any specified subset of the sample space is called an event.

Central subjects in probability theory include discrete and continuous random variables, probability distributions, and stochastic processes (which provide mathematical abstractions of non-deterministic or uncertain processes or measured quantities that may either be single occurrences or evolve over time in a random fashion). Although it is not possible to perfectly predict random events, much can be said about their behavior. Two major results in probability theory describing such behaviour are the law of large numbers and the central limit theorem.

As a mathematical foundation for statistics, probability theory is essential to many human activities that involve quantitative analysis of data. Methods of probability theory also apply to descriptions of complex systems given only partial knowledge of their state, as in statistical mechanics or sequential estimation. A great discovery of twentieth-century physics was the probabilistic nature of physical phenomena at atomic scales, described in quantum mechanics.

8.3.1 Treatment

Most introductions to probability theory treat discrete probability distributions and continuous probability distributions separately. The measure theory-based treatment of probability covers the discrete, continuous, a mix of the two, and more.

Motivation

Consider an experiment that can produce a number of outcomes. The set of all outcomes is called the *sample space* of the experiment. The *power set* of the sample space (or equivalently, the event space) is formed by considering all different collections of possible results. For example, rolling an honest die produces one of six possible results. One collection of possible results corresponds to getting an odd number. Thus, the subset {1,3,5} is an element of the power set of the sample space of dice rolls. These collections are called *events*. In this case, {1,3,5} is the event that the die falls on some odd number. If the results that actually occur fall in a given event, that event is said to have occurred.

Probability is a way of assigning every "event" a value between zero and one, with the requirement that the event made up of all possible results (in our example, the event {1,2,3,4,5,6}) be assigned a value of one. To qualify as a probability distribution, the assignment of values must satisfy the requirement that if you look at a collection of mutually exclusive events (events that contain no common results, e.g., the events {1,6}, {3}, and {2,4} are all mutually exclusive), the probability that any of these events occurs is given by the sum of the probabilities of the events.

The probability that any one of the events {1,6}, {3}, or {2,4} will occur is 5/6. This is the same as saying that the probability of event {1,2,3,4,6} is 5/6. This event encompasses the possibility of any number except five being rolled. The mutually exclusive event {5} has a probability of 1/6, and the event {1,2,3,4,5,6} has a probability of 1, that is, absolute certainty.

When doing calculations using the outcomes of an experiment, it is necessary that all those elementary events have a number assigned to them. This is done using a random variable. A random variable is a function that assigns to each elementary event in the sample space a real number. This function is usually denoted by a capital letter. In the case of a die, the assignment of a number to certain elementary events can be done using the identity function. This does not always work. For example, when flipping a coin the two possible outcomes are "heads" and "tails". In this example, the random variable X could assign to the outcome "heads" the number "0" (X (heads) = 0 $X(\text{heads}) = 0$) and to the outcome "tails" the number "1" (X (tails) = 1 $X(\text{tails}) = 1$).

Discrete Probability Distributions

Discrete probability theory deals with events that occur in countable sample spaces.

- **Examples:** Throwing dice, experiments with decks of cards, random walk, and tossing coins.
- **Classical Definition:** Initially the probability of an event to occur was defined as the number of cases favorable for the event, over the number of total outcomes possible in an equiprobable sample space.

For example, if the event is "occurrence of an even number when a dice is rolled", the probability is given by $\frac{3}{6} = \frac{1}{2}$, since 3 faces out of the 6 have even numbers and each face has the same probability of appearing.

8.3.2 Classical Probability Distributions

Certain random variables occur very often in probability theory because they well describe many natural or physical processes. Their distributions, therefore, have gained *special importance* in probability theory. Some fundamental *discrete distributions* are the discrete uniform, Bernoulli, binomial, negative binomial, Poisson and geometric distributions. Important *continuous distributions* include the continuous uniform, normal, exponential, gamma and beta distributions.

8.4 APPLICATIONS

Probability theory is applied in everyday life in risk assessment and modeling. The insurance industry and markets use actuarial science to determine pricing and make trading decisions. Governments apply probabilistic methods in environmental regulation, entitlement analysis, and financial regulation.

An example of the use of probability theory in equity trading is the effect of the perceived probability of any widespread Middle East conflict on oil prices, which have ripple effects in the economy as a whole. An assessment by a commodity trader that a war is more likely can send that commodity's prices up or down, and signals other traders of that opinion. Accordingly, the probabilities are neither assessed independently nor necessarily rationally. The theory of behavioral finance emerged to describe the effect of such groupthink on pricing, on policy, and on peace and conflict.

In addition to financial assessment, probability can be used to analyze trends in biology (e.g., disease spread) as well as ecology (e.g., biological Punnett squares). As with finance, risk assessment can be used as a statistical tool to calculate the likelihood of undesirable events occurring, and can assist with implementing protocols to avoid encountering such circumstances. Probability is used to design games of chance so that casinos can make a guaranteed profit, yet provide payouts to players that are frequent enough to encourage continued play.

Another significant application of probability theory in everyday life is reliability. Many consumer products, such as automobiles and consumer electronics, use reliability theory in product design to reduce the probability of failure. Failure probability may influence a manufacturer's decisions on a product's warranty.

The cache language model and other statistical language models that are used in natural language processing are also examples of applications of probability theory.

8.5 RELATION TO RANDOMNESS AND PROBABILITY IN QUANTUM MECHANICS

In a deterministic universe, based on Newtonian concepts, there would be no probability if all conditions were known (Laplace's demon) (but there are situations in which sensitivity to initial conditions exceeds our ability to measure them, i.e. know them). In the case of a roulette wheel, if the force of the hand and the period of that force are known, the number on which the ball will stop would be a certainty (though as a practical matter, this would likely be true only of a roulette wheel that had not been exactly levelled – as Thomas A. Bass' Newtonian Casino revealed). This also assumes knowledge of inertia and friction of the wheel, weight, smoothness, and roundness of the ball, variations in hand speed during the turning, and so forth. A probabilistic description can thus be more useful than Newtonian mechanics for analyzing the pattern of outcomes of repeated rolls of a roulette wheel. Physicists face the same situation in the kinetic theory of gases, where the system, while deterministic *in principle*, is so complex (with the number of molecules typically the order of magnitude of the Avogadro constant 6.02×10^{23}) that only a statistical description of its properties is feasible.

Probability theory is required to describe quantum phenomena. A revolutionary discovery of early 20th century physics was the random character of all physical processes that occur at sub-atomic scales and are governed by the laws of quantum mechanics. The objective wave function evolves deterministically but, according to the Copenhagen interpretation, it deals with probabilities of observing, the outcome being explained by a wave function collapse when an observation is made. However, the loss of determinism for the sake of instrumentalism did not meet with universal approval. Albert Einstein famously remarked in a letter to Max Born: "I am convinced that God does not play dice". Like Einstein, Erwin Schrödinger, who discovered the wave function, believed quantum mechanics is a statistical approximation of an underlying deterministic reality. In some modern interpretations of the statistical mechanics of measurement, quantum decoherence is invoked to account for the appearance of subjectively probabilistic experimental outcomes.

CHAPTER 9

PROBABILITY DISTRIBUTION

In probability theory and statistics, a probability distribution is the mathematical function that gives the probabilities of occurrence of possible outcomes for an experiment. It is a mathematical description of a random phenomenon in terms of its sample space and the probabilities of events (subsets of the sample space).

For instance, if X is used to denote the outcome of a coin toss ("the experiment"), then the probability distribution of X would take the value 0.5 (1 in 2 or 1/2) for X = heads, and 0.5 for X = tails (assuming that the coin is fair). More commonly, probability distributions are used to compare the relative occurrence of many different random values. Probability distributions can be defined in different ways and for discrete or for continuous variables. Distributions with special properties or for especially important applications are given specific names.

9.1 INTRODUCTION

A probability distribution is a mathematical description of the probabilities of events, subsets of the sample space. The sample space, often represented in notation by Ω is the set of all possible outcomes of a random phenomenon being observed. The sample space may be any set: a set of real numbers, a set of descriptive labels, a set of vectors, a set of arbitrary non-numerical values, etc. For example, the sample space of a coin flip could be Ω = { "heads", "tails" } .

To define probability distributions for the specific case of random variables (so the sample space can be seen as a numeric set), it is common to distinguish between discrete and absolutely continuous random variables. In the discrete case, it is sufficient to specify a probability mass function P assigning a probability to each possible outcome (e.g. when throwing a fair die, each of the six digits "1" to "6", corresponding to the number of dots on the die, has the probability $\frac{1}{6}$). The probability of an event is then defined to be the sum of the probabilities of all outcomes that satisfy the event; for example, the probability of the event "the die rolls an even value" is

$$p(2) + p(4) + p(6) = \tfrac{1}{6} + \tfrac{1}{6} + \tfrac{1}{6} = \tfrac{1}{2}.$$

In contrast, when a random variable takes values from a continuum then by convention, any individual outcome is assigned probability zero. For such continuous random variables, only events that include infinitely many outcomes such as intervals have probability greater than 0. For example, consider measuring the weight of a piece of ham in the supermarket, and assume the scale can provide arbitrarily many digits of precision. Then, the probability that it weighs *exactly* 500g must be zero because no matter how high the level of precision chosen, it cannot be assumed that there are no non-zero decimal digits in the remaining omitted digits ignored by the precision level.

However, for the same use case, it is possible to meet quality control requirements such as that a package of "500 g" of ham must weigh between 490 g and 510 g with at least 98% probability. This is possible because this measurement does not require as much precision from the underlying equipment.

Absolutely continuous probability distributions can be described in several ways. The probability density function describes the infinitesimal probability of any given value, and the probability that the outcome lies in a given interval can be computed by integrating the probability density function over that interval.

9.2 CUMULATIVE DISTRIBUTION FUNCTION

In the special case of a real-valued random variable, the probability distribution can equivalently be represented by a cumulative distribution function instead of a probability measure. The cumulative distribution function of a random variable X with regard to a probability distribution p p is defined as

$$F(x) = P(X \leq x).$$

Any probability distribution can be decomposed as the mixture of a discrete, an absolutely continuous and a singular continuous distribution, and thus any cumulative distribution function admits a decomposition as the convex sum of the three according cumulative distribution functions.

9.3 DISCRETE PROBABILITY DISTRIBUTION

A discrete probability distribution is the probability distribution of a random variable that can take on only a countable number of values (almost surely) which means that the probability of any event E E can be expressed as a (finite or countably infinite) sum:

$$P(X \in E) = \sum_{\omega \in A \cap E} P(X = \omega),$$

Well-known discrete probability distributions used in statistical modeling include the Poisson distribution, the Bernoulli distribution, the binomial distribution, the geometric distribution, the negative binomial distribution and categorical distribution. When a sample (a set of observations) is drawn from a larger population, the sample points have an empirical distribution that is discrete, and which provides information about the population distribution. Additionally, the discrete uniform distribution is commonly used in computer programs that make equal-probability random selections between a number of choices.

9.3.1 Cumulative Distribution Function

A real-valued discrete random variable can equivalently be defined as a random variable whose cumulative distribution function increases only by jump discontinuities—that is, its cdf increases only where it "jumps" to a higher value, and is constant in intervals without jumps.

The points where jumps occur are precisely the values which the random variable may take. Thus the cumulative distribution function has the form

$$F(x) = P(X \le x) = \sum_{\omega \le x} p(\omega).$$

The points where the cdf jumps always form a countable set; this may be any countable set and thus may even be dense in the real numbers.

9.4 COMMON PROBABILITY DISTRIBUTIONS AND THEIR APPLICATIONS

The concept of the probability distribution and the random variables which they describe underlies the mathematical discipline of probability theory, and the science of statistics.

There is spread or variability in almost any value that can be measured in a population (e.g. height of people, durability of a metal, sales growth, traffic flow, etc.); almost all measurements are made with some intrinsic error; in physics, many processes are described probabilistically, from the kinetic properties of gases to the quantum mechanical description of fundamental particles. For these and many other reasons, simple numbers are often inadequate for describing a quantity, while probability distributions are often more appropriate.

All of the univariate distributions below are singly peaked; that is, it is assumed that the values cluster around a single point. In practice, actually observed quantities may cluster around multiple values. Such quantities can be modeled using a mixture distribution.

9.4.1 Linear Growth (e.g. Errors, Offsets)

- Normal distribution (Gaussian distribution), for a single such quantity; the most commonly used absolutely continuous distribution

9.4.2 Exponential Growth (e.g. Prices, Incomes, Populations)

- Log-normal distribution, for a single such quantity whose log is normally distributed

- Pareto distribution, for a single such quantity whose log is exponentially distributed; the prototypical power law distribution

9.4.3 Uniformly Distributed Quantities

- Discrete uniform distribution, for a finite set of values (e.g. the outcome of a fair dice)
- Continuous uniform distribution, for absolutely continuously distributed values

9.4.4 Bernoulli Trials (Yes/No Events, with a Given Probability)

- **Basic Distributions:**
  - Bernoulli distribution, for the outcome of a single Bernoulli trial (e.g. success/failure, yes/no)
  - Binomial distribution, for the number of "positive occurrences" (e.g. successes, yes votes, etc.) given a fixed total number of independent occurrences
  - Negative binomial distribution, for binomial-type observations but where the quantity of interest is the number of failures before a given number of successes occurs
  - Geometric distribution, for binomial-type observations but where the quantity of interest is the number of failures before the first success; a special case of the negative binomial distribution
- **Related to Sampling Schemes Over a Finite Population:**
  - Hypergeometric distribution, for the number of "positive occurrences" (e.g. successes, yes votes, etc.) given a fixed number of total occurrences, using sampling without replacement
  - Beta-binomial distribution, for the number of "positive occurrences" (e.g. successes, yes votes, etc.) given a fixed number of total occurrences, sampling using a Pólya urn model (in some sense, the "opposite" of sampling without replacement)

9.4.5 Categorical Outcomes (Events with K Possible Outcomes)

- Categorical distribution, for a single categorical outcome (e.g. yes/no/maybe in a survey); a generalization of the Bernoulli distribution
- Multinomial distribution, for the number of each type of categorical outcome, given a fixed number of total outcomes; a generalization of the binomial distribution
- Multivariate hypergeometric distribution, similar to the multinomial distribution, but using sampling without replacement; a generalization of the hypergeometric distribution

9.4.6 Poisson Process (Events that Occur Independently with a Given Rate)

- Poisson distribution, for the number of occurrences of a Poisson-type event in a given period of time
- Exponential distribution, for the time before the next Poisson-type event occurs
- Gamma distribution, for the time before the next k Poisson-type events occur

9.4.7 Absolute Values of Vectors with Normally Distributed Components

- Rayleigh distribution, for the distribution of vector magnitudes with Gaussian distributed orthogonal components. Rayleigh distributions are found in RF signals with Gaussian real and imaginary components.
- Rice distribution, a generalization of the Rayleigh distributions for where there is a stationary background signal component. Found in Rician fading of radio signals due to multipath propagation and in MR images with noise corruption on non-zero NMR signals.

9.4.8 Normally Distributed Quantities Operated with Sum of Squares

- Chi-squared distribution, the distribution of a sum of squared standard normal variables; useful e.g. for inference regarding the sample variance of normally distributed samples.
- Student's t distribution, the distribution of the ratio of a standard normal variable and the square root of a scaled chi squared variable; useful for inference regarding the mean of normally distributed samples with unknown variance.
- F-distribution, the distribution of the ratio of two scaled chi squared variables; useful e.g. for inferences that involve comparing variances or involving R-squared (the squared correlation coefficient).

9.4.9 As Conjugate Prior Distributions in Bayesian Inference

- Beta distribution, for a single probability (real number between 0 and 1); conjugate to the Bernoulli distribution and binomial distribution
- Gamma distribution, for a non-negative scaling parameter; conjugate to the rate parameter of a Poisson distribution or exponential distribution, the precision (inverse variance) of a normal distribution, etc.
- Dirichlet distribution, for a vector of probabilities that must sum to 1;

conjugate to the categorical distribution and multinomial distribution; generalization of the beta distribution

- Wishart distribution, for a symmetric non-negative definite matrix; conjugate to the inverse of the covariance matrix of a multivariate normal distribution; generalization of the gamma distribution

9.4.10 Some Specialized Applications of Probability Distributions

- The cache language models and other statistical language models used in natural language processing to assign probabilities to the occurrence of particular words and word sequences do so by means of probability distributions.
- In quantum mechanics, the probability density of finding the particle at a given point is proportional to the square of the magnitude of the particle's wavefunction at that point. Therefore, the probability distribution function of the position of a particle is described by $P_{a\le x\le b}(t)=\int_a^b dx|\Psi(x,t)|^2$, probability that the particle's position x will be in the interval $a \le x \le b$ in dimension one, and a similar triple integral in dimension three. This is a key principle of quantum mechanics.
- Probabilistic load flow in power-flow study explains the uncertainties of input variables as probability distribution and provides the power flow calculation also in term of probability distribution.
- Prediction of natural phenomena occurrences based on previous frequency distributions such as tropical cyclones, hail, time in between events, etc.

9.5 FITTING

Probability distribution fitting or simply distribution fitting is the fitting of a probability distribution to a series of data concerning the repeated measurement of a variable phenomenon. The aim of distribution fitting is to predict the probability or to forecast the frequency of occurrence of the magnitude of the phenomenon in a certain interval.

There are many probability distributions of which some can be fitted more closely to the observed frequency of the data than others, depending on the characteristics of the phenomenon and of the distribution. The distribution giving a close fit is supposed to lead to good predictions. In distribution fitting, therefore, one needs to select a distribution that suits the data well.

CHAPTER 10
SAMPLING

In statistics, quality assurance, and survey methodology, sampling is the selection of a subset or a statistical sample (termed sample for short) of individuals from within a statistical population to estimate characteristics of the whole population. The subset is meant to reflect the whole population and statisticians attempt to collect samples that are representative of the population. Sampling has lower costs and faster data collection compared to recording data from the entire population, and thus, it can provide insights in cases where it is infeasible to measure an entire population.

Each observation measures one or more properties (such as weight, location, colour or mass) of independent objects or individuals. In survey sampling, weights can be applied to the data to adjust for the sample design, particularly in stratified sampling. Results from probability theory and statistical theory are employed to guide the practice. In business and medical research, sampling is widely used for gathering information about a population. Acceptance sampling is used to determine if a production lot of material meets the governing specifications.

10.1 HISTORY

Random sampling by using lots is an old idea, mentioned several times in the Bible. In 1786, Pierre Simon Laplace estimated the population of France by using a sample, along with ratio estimator. He also computed probabilistic estimates of the error. These were not expressed as modern confidence intervals but as the sample size that would be needed to achieve a particular upper bound on the sampling error with probability 1000/1001.

His estimates used Bayes' theorem with a uniform prior probability and assumed that his sample was random. Alexander Ivanovich Chuprov introduced sample surveys to Imperial Russia in the 1870s.

In the US, the 1936 *Literary Digest* prediction of a Republican win in the presidential election went badly awry, due to severe bias. More than two million people responded to the study with their names obtained through magazine subscription lists and telephone directories. It was not appreciated that these lists were heavily biased towards Republicans and the resulting sample, though very large, was deeply flawed.

Elections in Singapore have adopted this practice since the 2015 election, also known as the sample counts, whereas according to the Elections Department (ELD), their country's election commission, sample counts help reduce speculation and misinformation, while helping election officials to check against the election result for that electoral division. The reported sample counts yield a fairly accurate indicative result with a 95% confidence interval at a margin of error within 4-5%; ELD reminded the public that sample counts are separate from official results, and only the returning officer will declare the official results once vote counting is complete.

10.2 POPULATION DEFINITION

Successful statistical practice is based on focused problem definition. In sampling, this includes defining the "population" from which our sample is drawn. A population can be defined as including all people or items with the characteristics one wishes to understand. Because there is very rarely enough time or money to gather information from everyone or everything in a population, the goal becomes finding a representative sample (or subset) of that population.

Sometimes what defines a population is obvious. For example, a manufacturer needs to decide whether a batch of material from production is of high enough quality to be released to the customer or should be scrapped or reworked due to poor quality. In this case, the batch is the population.

Although the population of interest often consists of physical objects, sometimes it is necessary to sample over time, space, or some combination of these dimensions. For instance, an investigation of supermarket staffing could examine checkout line length at various times, or a study on endangered penguins might aim to understand their usage of various hunting grounds over time. For the time dimension, the focus may be on periods or discrete occasions.

In other cases, the examined 'population' may be even less tangible. For example, Joseph Jagger studied the behaviour of roulette wheels at a casino in Monte Carlo, and used this to identify a biased wheel. In this case, the 'population' Jagger wanted to investigate was the overall behaviour of the wheel (i.e. the probability distribution of its results over infinitely many trials), while his 'sample' was formed from observed results from that wheel. Similar considerations arise when taking repeated measurements of properties of materials such as the electrical conductivity of copper.

This situation often arises when seeking knowledge about the cause system of which the *observed* population is an outcome. In such cases, sampling theory may treat the observed population as a sample from a larger 'superpopulation'. For example, a researcher might study the success rate of a new 'quit smoking' program on a test group of 100 patients, in order to predict the effects of the program if it were made available nationwide. Here the superpopulation is "everybody in the country, given access to this treatment" – a group that does not yet exist since the program is not yet available to all.

The population from which the sample is drawn may not be the same as the population from which information is desired. Often there is a large but not complete overlap between these two groups due to frame issues etc. Sometimes they may be entirely separate – for instance, one might study rats in order to get a better understanding of human health, or one might study records from people born in 2008 in order to make predictions about people born in 2009.

Time spent in making the sampled population and population of concern precise is often well spent because it raises many issues, ambiguities, and questions that would otherwise have been overlooked at this stage.

10.3 SAMPLING FRAME

In the most straightforward case, such as the sampling of a batch of material from production (acceptance sampling by lots), it would be most desirable to identify and measure every single item in the population and to include any one of them in our sample.

However, in the more general case this is not usually possible or practical. There is no way to identify all rats in the set of all rats. Where voting is not compulsory, there is no way to identify which people will vote at a forthcoming election (in advance of the election). These imprecise populations are not amenable to sampling in any of the ways below and to which we could apply statistical theory.

As a remedy, we seek a sampling frame which has the property that we can identify every single element and include any in our sample. The most straightforward type of frame is a list of elements of the population (preferably the entire population) with appropriate contact information.

For example, in an opinion poll, possible sampling frames include an electoral register and a telephone directory.

A probability sample is a sample in which every unit in the population has a chance (greater than zero) of being selected in the sample, and this probability can be accurately determined. The combination of these traits makes it possible to produce unbiased estimates of population totals, by weighting sampled units according to their probability of selection.

- **Example:** We want to estimate the total income of adults living in a given street. We visit each household in that street, identify all adults living there, and randomly select one adult from each household. (For example, we can allocate each person a random number, generated from a *uniform distribution* between 0 and 1, and select the person with the highest number in each household). We then interview the selected person and find their income.

- People living on their own are certain to be selected, so we simply add their income to our estimate of the total. But a person living in a household of two adults has only a one-in-two chance of selection. To reflect this, when we come to such a household, we would count the selected person's income twice towards the total. (The person who is selected from that household can be loosely viewed as also representing the person who isn't selected.)

In the above example, not everybody has the same probability of selection; what makes it a probability sample is the fact that each person's probability is known. When every element in the population *does* have the same probability of selection, this is known as an 'equal probability of selection' (EPS) design. Such designs are also referred to as 'self-weighting' because all sampled units are given the same weight.

Probability sampling includes: simple random sampling, systematic sampling, stratified sampling, probability-proportional-to-size sampling, and cluster or multistage sampling. These various ways of probability sampling have two things in common:

- Every element has a known nonzero probability of being sampled.
- Involves random selection at some point.

10.3.1 Nonprobability Sampling

Nonprobability sampling is any sampling method where some elements of the population have *no* chance of selection (these are sometimes referred to as 'out of coverage'/'undercovered'), or where the probability of selection cannot be accurately determined. It involves the selection of elements based on assumptions regarding the population of interest, which forms the criteria for selection. Hence, because the selection of elements is nonrandom, nonprobability sampling does not allow the estimation of sampling errors. These conditions give rise to exclusion bias, placing limits on how much information a sample can provide about the population. Information about the relationship between sample and population is limited, making it difficult to extrapolate from the sample to the population.

- **Example:** We visit every household in a given street, and interview the first person to answer the door. In any household with more than one occupant, this is a nonprobability sample, because some people are more likely to answer the door (e.g. an unemployed person who spends most of their time at home is more likely to answer than an employed housemate who might be at work when the interviewer calls) and it's not practical to calculate these probabilities.

Nonprobability sampling methods include convenience sampling, quota sampling, and purposive sampling. In addition, nonresponse effects may turn *any* probability design into a nonprobability design if the characteristics of nonresponse are not well understood, since nonresponse effectively modifies each element's probability of being sampled.

10.4 SAMPLING METHODS

Within any of the types of frames identified above, a variety of sampling methods can be employed individually or in combination. Factors commonly influencing the choice between these designs include:

- Nature and quality of the frame.
- Availability of auxiliary information about units on the frame.
- Accuracy requirements, and the need to measure accuracy.
- Whether detailed analysis of the sample is expected.
- Cost/operational concerns.

10.4.1 Simple Random Sampling

In a simple random sample (SRS) of a given size, all subsets of a sampling frame have an equal probability of being selected. Each element of the frame thus has an equal probability of selection: the frame is not subdivided or partitioned. Furthermore, any given *pair* of elements has the same chance of selection as any other such pair (and similarly for triples, and so on). This minimizes bias and simplifies analysis of results. In particular, the variance between individual results within the sample is a good indicator of variance in the overall population, which makes it relatively easy to estimate the accuracy of results.

Simple random sampling can be vulnerable to sampling error because the randomness of the selection may result in a sample that does not reflect the makeup of the population. For instance, a simple random sample of ten people from a given country will *on average* produce five men and five women, but any given trial is likely to over represent one sex and underrepresent the other. Systematic and stratified techniques attempt to overcome this problem by "using information about the population" to choose a more "representative" sample.

Also, simple random sampling can be cumbersome and tedious when sampling from a large target population. In some cases, investigators are interested in research questions specific to subgroups of the population. For example, researchers might be interested in examining whether cognitive ability as a predictor of job performance is equally applicable across racial groups. Simple random sampling cannot accommodate the needs of researchers in this situation, because it does not provide subsamples of the population, and other sampling strategies, such as stratified sampling, can be used instead.

10.4.2 Systematic Sampling

Systematic sampling (also known as interval sampling) relies on arranging the study population according to some ordering scheme and then selecting elements at regular intervals through that ordered list. Systematic sampling involves a random start and then proceeds with the selection of every *k*th element from then onwards. In this case, *k*=(population size/sample size). It is important that the starting point is not automatically the first in the list, but is instead randomly chosen from within the first to the *k*th element in the list. A simple example would be to select every 10th name from the telephone directory (an 'every 10th' sample, also referred to as 'sampling with a skip of 10').

As long as the starting point is randomized, systematic sampling is a type of probability sampling. It is easy to implement and the stratification induced can make it efficient, *if* the variable by which the list is ordered is correlated with the variable of interest. 'Every 10th' sampling is especially useful for efficient sampling from databases.

For example, suppose we wish to sample people from a long street that starts in a poor area (house No. 1) and ends in an expensive district (house No. 1000). A simple random selection of addresses from this street could easily end up with too many from the high end and too few from the low end (or vice versa), leading to an unrepresentative sample. Selecting (e.g.) every 10th street number along the street ensures that the sample is spread evenly along the length of the street, representing all of these districts. (If we always start at house #1 and end at #991, the sample is slightly biased towards the low end; by randomly selecting the start between #1 and #10, this bias is eliminated.)

However, systematic sampling is especially vulnerable to periodicities in the list. If periodicity is present and the period is a multiple or factor of the interval used, the sample is especially likely to be *un*representative of the overall population, making the scheme less accurate than simple random sampling.

For example, consider a street where the odd-numbered houses are all on the north (expensive) side of the road, and the even-numbered houses are all on the south (cheap) side. Under the sampling scheme given above, it is impossible to get a representative sample; either the houses sampled will *all* be from the odd-numbered, expensive side, or they will *all* be from the even-numbered, cheap side, unless the researcher has previous knowledge of this bias and avoids it by a using a skip which ensures jumping between the two sides (any odd-numbered skip).

Another drawback of systematic sampling is that even in scenarios where it is more accurate than SRS, its theoretical properties make it difficult to *quantify* that accuracy. (In the two examples of systematic sampling that are given above, much of the potential sampling error is due to variation between neighbouring houses – but because this method never selects two neighbouring houses, the sample will not give us any information on that variation.)

As described above, systematic sampling is an EPS method, because all elements have the same probability of selection (in the example given, one in ten). It is *not* 'simple random sampling' because different subsets of the same size have different selection probabilities – e.g. the set {4,14,24,...,994} has a one-in-ten probability of selection, but the set {4,13,24,34,...} has zero probability of selection.

10.4.3 Stratified Sampling

When the population embraces a number of distinct categories, the frame can be organized by these categories into separate "strata." Each stratum is then sampled as an independent sub-population, out of which individual elements can be randomly selected. The ratio of the size of this random selection (or sample) to the size of the population is called a sampling fraction. There are several potential benefits to stratified sampling.

First, dividing the population into distinct, independent strata can enable researchers to draw inferences about specific subgroups that may be lost in a more generalized random sample.

Second, utilizing a stratified sampling method can lead to more efficient statistical estimates (provided that strata are selected based upon relevance to the criterion in question, instead of availability of the samples). Even if a stratified sampling approach does not lead to increased statistical efficiency, such a tactic will not result in less efficiency than would simple random sampling, provided that each stratum is proportional to the group's size in the population.

Third, it is sometimes the case that data are more readily available for individual, pre-existing strata within a population than for the overall population; in such cases, using a stratified sampling approach may be more convenient than aggregating data across groups (though this may potentially be at odds with the previously noted importance of utilizing criterion-relevant strata).

Finally, since each stratum is treated as an independent population, different sampling approaches can be applied to different strata, potentially enabling researchers to use the approach best suited (or most cost-effective) for each identified subgroup within the population.

There are, however, some potential drawbacks to using stratified sampling. First, identifying strata and implementing such an approach can increase the cost and complexity of sample selection, as well as leading to increased complexity of population estimates. Second, when examining multiple criteria, stratifying variables may be related to some, but not to others, further complicating the design, and potentially reducing the utility of the strata.

Finally, in some cases (such as designs with a large number of strata, or those with a specified minimum sample size per group), stratified sampling can potentially require a larger sample than would other methods (although in most cases, the required sample size would be no larger than would be required for simple random sampling).

A stratified sampling approach is most effective when three conditions are met:

- Variability within strata are minimized.
- Variability between strata are maximized.
- The variables upon which the population is stratified are strongly correlated with the desired dependent variable.

Advantages over other sampling methods:

- Focuses on important subpopulations and ignores irrelevant ones.
- Allows use of different sampling techniques for different subpopulations.
- Improves the accuracy/efficiency of estimation.
- Permits greater balancing of statistical power of tests of differences between strata by sampling equal numbers from strata varying widely in size.

Disadvantages:

- Requires selection of relevant stratification variables which can be difficult.
- Is not useful when there are no homogeneous subgroups.
- Can be expensive to implement.

Poststratification

Stratification is sometimes introduced after the sampling phase in a process called "poststratification". This approach is typically implemented due to a lack of prior knowledge of an appropriate stratifying variable or when the experimenter lacks the necessary information to create a stratifying variable during the sampling phase. Although the method is susceptible to the pitfalls of post hoc approaches, it can provide several benefits in the right situation. Implementation usually follows a simple random sample. In addition to allowing for stratification on an ancillary variable, poststratification can be used to implement weighting, which can improve the precision of a sample's estimates.

Oversampling

Choice-based sampling is one of the stratified sampling strategies. In choice-based sampling, the data are stratified on the target and a sample is taken from each stratum so that the rare target class will be more represented in the sample. The model is then built on this biased sample. The effects of the input variables on the target are often estimated with more precision with the choice-based sample even when a smaller overall sample size is taken, compared to a random sample. The results usually must be adjusted to correct for the oversampling.

10.4.4 Probability-Proportional-to-Size Sampling

In some cases the sample designer has access to an "auxiliary variable" or "size measure", believed to be correlated to the variable of interest, for each element in the population. These data can be used to improve accuracy in sample design. One option is to use the auxiliary variable as a basis for stratification, as discussed above.

Another option is probability proportional to size ('PPS') sampling, in which the selection probability for each element is set to be proportional to its size measure, up to a maximum of 1. In a simple PPS design, these selection probabilities can then be used as the basis for Poisson sampling. However, this has the drawback of variable sample size, and different portions of the population may still be over- or under-represented due to chance variation in selections.

Systematic sampling theory can be used to create a probability proportionate to size sample. This is done by treating each count within the size variable as a single sampling unit. Samples are then identified by selecting at even intervals among these counts within the size variable. This method is sometimes called PPS-sequential or monetary unit sampling in the case of audits or forensic sampling.

- **Example:** Suppose we have six schools with populations of 150, 180, 200, 220, 260, and 490 students respectively (total 1500 students), and we want to use student population as the basis for a PPS sample of size three. To do this, we could allocate the first school numbers 1 to 150, the second school 151 to 330 (= 150 + 180), the third school 331 to 530, and so on to the last school (1011 to 1500). We then generate a random start between 1 and 500 (equal to 1500/3) and count through the school populations by multiples of 500. If our random start was 137, we would select the schools which have been allocated numbers 137, 637, and 1137, i.e. the first, fourth, and sixth schools.

The PPS approach can improve accuracy for a given sample size by concentrating sample on large elements that have the greatest impact on population estimates. PPS sampling is commonly used for surveys of businesses, where element size varies greatly and auxiliary information is often available – for instance, a survey attempting to measure the number of guest-nights spent in hotels might use each hotel's number of rooms as an auxiliary variable. In some cases, an older measurement of the variable of interest can be used as an auxiliary variable when attempting to produce more current estimates.

10.4.5 Cluster Sampling

Sometimes it is more cost-effective to select respondents in groups ('clusters'). Sampling is often clustered by geography, or by time periods. (Nearly all samples are in some sense 'clustered' in time – although this is rarely taken into account in the analysis.) For instance, if surveying households within a city, we might choose to select 100 city blocks and then interview every household within the selected blocks.

Clustering can reduce travel and administrative costs. In the example above, an interviewer can make a single trip to visit several households in one block, rather than having to drive to a different block for each household.

It also means that one does not need a sampling frame listing all elements in the target population. Instead, clusters can be chosen from a cluster-level frame, with an element-level frame created only for the selected clusters. In the example above, the sample only requires a block-level city map for initial selections, and then a household-level map of the 100 selected blocks, rather than a household-level map of the whole city.

Cluster sampling (also known as clustered sampling) generally increases the variability of sample estimates above that of simple random sampling, depending on how the clusters differ between one another as compared to the within-cluster variation. For this reason, cluster sampling requires a larger sample than SRS to achieve the same level of accuracy – but cost savings from clustering might still make this a cheaper option.

Cluster sampling is commonly implemented as multistage sampling. This is a complex form of cluster sampling in which two or more levels of units are embedded one in the other. The first stage consists of constructing the clusters that will be used to sample from. In the second stage, a sample of primary units is randomly selected from each cluster (rather than using all units contained in all selected clusters). In following stages, in each of those selected clusters, additional samples of units are selected, and so on. All ultimate units (individuals, for instance) selected at the last step of this procedure are then surveyed. This technique, thus, is essentially the process of taking random subsamples of preceding random samples.

Multistage sampling can substantially reduce sampling costs, where the complete population list would need to be constructed (before other sampling methods could be applied). By eliminating the work involved in describing clusters that are not selected, multistage sampling can reduce the large costs associated with traditional cluster sampling. However, each sample may not be a full representative of the whole population.

10.4.6 Quota Sampling

In quota sampling, the population is first segmented into mutually exclusive sub-groups, just as in stratified sampling. Then judgement is used to select the subjects or units from each segment based on a specified proportion. For example, an interviewer may be told to sample 200 females and 300 males between the age of 45 and 60.

It is this second step which makes the technique one of non-probability sampling. In quota sampling the selection of the sample is non-random. For example, interviewers might be tempted to interview those who look most helpful. The problem is that these samples may be biased because not everyone gets a chance of selection. This random element is its greatest weakness and quota versus probability has been a matter of controversy for several years.

10.4.7 Minimax Sampling

In imbalanced datasets, where the sampling ratio does not follow the population statistics, one can resample the dataset in a conservative manner called minimax sampling. The minimax sampling has its origin in Anderson minimax ratio whose value is proved to be 0.5: in a binary classification, the class-sample sizes should be chosen equally. This ratio can be proved to be minimax ratio only under the assumption of LDA classifier with Gaussian distributions. The notion of minimax sampling is recently developed for a general class of classification rules, called class-wise smart classifiers. In this case, the sampling ratio of classes is selected so that the worst case classifier error over all the possible population statistics for class prior probabilities, would be the best.

10.4.8 Accidental Sampling

Accidental sampling (sometimes known as grab, convenience or opportunity sampling) is a type of nonprobability sampling which involves the sample being drawn from that part of the population which is close to hand. That is, a population is selected because it is readily available and convenient. It may be through meeting the person or including a person in the sample when one meets them or chosen by finding them through technological means such as the internet or through phone.

The researcher using such a sample cannot scientifically make generalizations about the total population from this sample because it would not be representative enough. For example, if the interviewer were to conduct such a survey at a shopping center early in the morning on a given day, the people that they could interview would be limited to those given there at that given time, which would not represent the views of other members of society in such an area, if the survey were to be conducted at different times of day and several times per week. This type of sampling is most useful for pilot testing. Several important considerations for researchers using convenience samples include:

- Are there controls within the research design or experiment which can serve to lessen the impact of a non-random convenience sample, thereby ensuring the results will be more representative of the population?
- Is there good reason to believe that a particular convenience sample would or should respond or behave differently than a random sample from the same population?
- Is the question being asked by the research one that can adequately be answered using a convenience sample?

In social science research, snowball sampling is a similar technique, where existing study subjects are used to recruit more subjects into the sample. Some variants of snowball sampling, such as respondent driven sampling, allow calculation of selection probabilities and are probability sampling methods under certain conditions.

10.4.9 Voluntary Sampling

The voluntary sampling method is a type of non-probability sampling. Volunteers choose to complete a survey. Volunteers may be invited through advertisements in social media. The target population for advertisements can be selected by characteristics like location, age, sex, income, occupation, education, or interests using tools provided by the social medium. The advertisement may include a message about the research and link to a survey. After following the link and completing the survey, the volunteer submits the data to be included in the sample population. This method can reach a global population but is limited by the campaign budget. Volunteers outside the invited population may also be included in the sample. It is difficult to make generalizations from this sample because it may not represent the total population. Often, volunteers have a strong interest in the main topic of the survey.

10.4.10 Line-Intercept Sampling

Line-intercept sampling is a method of sampling elements in a region whereby an element is sampled if a chosen line segment, called a "transect", intersects the element.

10.4.11 Panel Sampling

Panel sampling is the method of first selecting a group of participants through a random sampling method and then asking that group for (potentially the same) information several times over a period of time. Therefore, each participant is interviewed at two or more time points; each period of data collection is called a "wave". The method was developed by sociologist Paul Lazarsfeld in 1938 as a means of studying political campaigns. This longitudinal sampling-method allows estimates of changes in the population, for example with regard to chronic illness to job stress to weekly food expenditures. Panel sampling can also be used to inform researchers about within-person health changes due to age or to help explain changes in continuous dependent variables such as spousal interaction. There have been several proposed methods of analyzing panel data, including MANOVA, growth curves, and structural equation modeling with lagged effects.

10.4.12 Snowball Sampling

Snowball sampling involves finding a small group of initial respondents and using them to recruit more respondents. It is particularly useful in cases where the population is hidden or difficult to enumerate.

10.4.13 Theoretical Sampling

Theoretical sampling occurs when samples are selected on the basis of the results of the data collected so far with a goal of developing a deeper understanding of the area or develop theories. Extreme or very specific cases might be selected in order to maximize the likelihood a phenomenon will actually be observable.

10.4.14 Active Sampling

In active sampling, the samples which are used for training a machine learning algorithm are actively selected, also compare active learning (machine learning).

10.5 REPLACEMENT OF SELECTED UNITS

Sampling schemes may be *without replacement* ('WOR' – no element can be selected more than once in the same sample) or *with replacement* ('WR' – an element may appear multiple times in the one sample).

For example, if we catch fish, measure them, and immediately return them to the water before continuing with the sample, this is a WR design, because we might end up catching and measuring the same fish more than once. However, if we do not return the fish to the water or tag and release each fish after catching it, this becomes a WOR design.

10.6 SAMPLE SIZE DETERMINATION

Formulas, tables, and power function charts are well known approaches to determine sample size. Steps for using sample size tables:

- Postulate the effect size of interest, α, and β.
- Check sample size table
  - Select the table corresponding to the selected α
  - Locate the row corresponding to the desired power
  - Locate the column corresponding to the estimated effect size.
  - The intersection of the column and row is the minimum sample size required.

10.7 SAMPLING AND DATA COLLECTION

Good data collection involves:

- Following the defined sampling process.
- Keeping the data in time order.
- Noting comments and other contextual events.
- Recording non-responses.

10.8 APPLICATIONS OF SAMPLING

Sampling enables the selection of right data points from within the larger data set to estimate the characteristics of the whole population. For example, there are about 600 million tweets produced every day. It is not necessary to look at all of them to determine the topics that are discussed during the day, nor is it necessary to look at all the tweets to determine the sentiment on each of the topics. A theoretical formulation for sampling Twitter data has been developed.

In manufacturing different types of sensory data such as acoustics, vibration, pressure, current, voltage, and controller data are available at short time intervals. To predict down-time it may not be necessary to look at all the data but a sample may be sufficient.

10.9 ERRORS IN SAMPLE SURVEYS

Survey results are typically subject to some error. Total errors can be classified into sampling errors and non-sampling errors. The term "error" here includes systematic biases as well as random errors.

10.9.1 Sampling Errors and Biases

Sampling errors and biases are induced by the sample design. They include:

- **Selection Bias:** When the true selection probabilities differ from those assumed in calculating the results.

- **Random Sampling Error:** Random variation in the results due to the elements in the sample being selected at random.

10.9.2 Non-Sampling Error

Non-sampling errors are other errors which can impact final survey estimates, caused by problems in data collection, processing, or sample design. Such errors may include:

- **Over-Coverage:** Inclusion of data from outside of the population
- **Under-Coverage:** Sampling frame does not include elements in the population.
- **Measurement Error:** E.g. when respondents misunderstand a question, or find it difficult to answer
- **Processing Error:** Mistakes in data coding
- **Non-Response or Participation Bias:** Failure to obtain complete data from all selected individuals

After sampling, a review is held of the exact process followed in sampling, rather than that intended, in order to study any effects that any divergences might have on subsequent analysis.

A particular problem involves *non-response*. Two major types of non-response exist:

- Unit nonresponse (lack of completion of any part of the survey),
- Item non-response (submission or participation in survey but failing to complete one or more components/questions of the survey).

In survey sampling, many of the individuals identified as part of the sample may be unwilling to participate, not have the time to participate (opportunity cost), or survey administrators may not have been able to contact them. In this case, there is a risk of differences between respondents and nonrespondents, leading to biased estimates of population parameters.

This is often addressed by improving survey design, offering incentives, and conducting follow-up studies which make a repeated attempt to contact the unresponsive and to characterize their similarities and differences with the rest of the frame.

The effects can also be mitigated by weighting the data (when population benchmarks are available) or by imputing data based on answers to other questions. Nonresponse is particularly a problem in internet sampling. Reasons for this problem may include improperly designed surveys, over-surveying (or survey fatigue), and the fact that potential participants may have multiple e-mail addresses, which they do not use anymore or do not check regularly.

10.10 SURVEY WEIGHTS

In many situations, the sample fraction may be varied by stratum and data will have to be weighted to correctly represent the population. Thus for example, a simple random sample of individuals in the United Kingdom might not include some in remote Scottish islands who would be inordinately expensive to sample. A cheaper method would be to use a stratified sample with urban and rural strata. The rural sample could be under-represented in the sample, but weighted up appropriately in the analysis to compensate.

More generally, data should usually be weighted if the sample design does not give each individual an equal chance of being selected. For instance, when households have equal selection probabilities but one person is interviewed from within each household, this gives people from large households a smaller chance of being interviewed. This can be accounted for using survey weights. Similarly, households with more than one telephone line have a greater chance of being selected in a random digit dialing sample, and weights can adjust for this. Weights can also serve other purposes, such as helping to correct for non-response.

10.11 METHODS OF PRODUCING RANDOM SAMPLES

- Random number table.
- Mathematical algorithms for pseudo-random number generators.
- Physical randomization devices such as coins, playing cards or sophisticated devices such as ERNIE.

CHAPTER 11

BASIS OF STATISTICAL INFERENCE

Statistical inference is the process of drawing conclusions about a population based on information gathered from a sample. It forms the backbone of decision-making in various scientific, medical, and business domains by using data to make generalizations or predictions. The ability to infer from a limited dataset allows researchers and practitioners to understand patterns, relationships, and potential future outcomes without needing to examine every member of a population.

At the heart of statistical inference are concepts such as estimation, hypothesis testing, confidence intervals, and p-values. These tools enable us to assess whether the results observed in a sample can be generalized to the larger population.

11.1 TYPES OF DATA IN STATISTICAL INFERENCE

Before delving into the methods of inference, it is important to understand the types of data used in statistical analysis. Data can be broadly categorized into two types: **qualitative (categorical)** and **quantitative (numerical)**.

11.1.1 Qualitative Data

Qualitative data represents categories or groups, such as gender, race, or disease status. These are non-numerical in nature and are often classified as nominal (unordered categories) or ordinal (ordered categories). For example, blood types (A, B, AB, O) are nominal, while pain levels (low, medium, high) are ordinal.

11.1.2 Quantitative Data

Quantitative data refers to numerical values, such as age, weight, or blood pressure. These can be classified as continuous (measurable and can take any value within a range) or discrete (countable and takes specific values). For instance, height is a continuous variable, while the number of children in a family is a discrete variable. The type of data influences the choice of statistical methods used for inference. Different tests and models are applicable depending on whether the data is categorical or numerical.

11.2 ASSUMPTIONS IN STATISTICAL INFERENCE

Statistical inference relies on several underlying assumptions that must be met for the results to be valid. Violating these assumptions can lead to incorrect conclusions, making it essential to ensure that the chosen methods are appropriate for the data at hand.

11.2.1 Random Sampling

One of the key assumptions in statistical inference is that the data used in analysis comes from a random sample of the population. This ensures that each member of the population has an equal chance of being included, which reduces bias and makes the sample representative of the population.

11.2.2 Independence of Observations

Another assumption is that the observations in the dataset are independent of one another. This means that the value of one observation does not influence or depend on the value of another observation. For instance, in medical trials, the health outcome of one patient should not affect the health outcome of another patient.

11.2.3 Normality

Many statistical methods, especially those involving parametric tests like t-tests and ANOVA, assume that the data follows a normal distribution. This assumption is critical for ensuring the validity of p-values and confidence intervals. In cases where the data is not normally distributed, non-parametric methods or data transformations may be needed.

11.3 KEY CONCEPTS IN STATISTICAL INFERENCE

There are several key concepts that form the basis of statistical inference, including point estimation, interval estimation, and hypothesis testing. These methods allow researchers to make inferences about population parameters based on sample data.

11.3.1 Point Estimation

In statistics, point estimation involves the use of sample data to calculate a single value (known as a point estimate since it identifies a point in some parameter space) which is to serve as a "best guess" or "best estimate" of an unknown population parameter (for example, the population mean). More formally, it is the application of a point estimator to the data to obtain a point estimate.

Point estimation can be contrasted with interval estimation: such interval estimates are typically either confidence intervals, in the case of frequentist inference, or credible intervals, in the case of Bayesian inference. More generally, a point estimator can be contrasted with a set estimator. Examples are given by confidence sets or credible sets. A point estimator can also be contrasted with a distribution estimator. Examples are given by confidence distributions, randomized estimators, and Bayesian posteriors.

Properties of Point Estimates

Biasedness

"Bias" is defined as the difference between the expected value of the estimator and the true value of the population parameter being estimated. It can also be described that the closer the expected value of a parameter is to the measured parameter, the lesser the bias.

When the estimated number and the true value is equal, the estimator is considered unbiased. This is called an *unbiased estimator.* The estimator will become a *best unbiased estimator* if it has minimum variance. However, a biased estimator with a small variance may be more useful than an unbiased estimator with a large variance. Most importantly, we prefer point estimators that have the smallest mean square errors.

If we let $T = h(X_1, X_2, \ldots, X_n)$ be an estimator based on a random sample $X_1, X_2, \ldots, X_n$, the estimator T is called an unbiased estimator for the parameter θ if $E[T] = \theta$, irrespective of the value of θ. For example, from the same random sample we have $E(\bar{x}) = \mu$ (mean) and $E(s^2) = \sigma^2$ (variance), then $\bar{x}$ and s^2 would be unbiased estimators for μ and σ^2. The difference $E[T] - \theta$ is called the bias of T; if this difference is nonzero, then T is called biased.

Consistency

Consistency is about whether the point estimate stays close to the value when the parameter increases its size. The larger the sample size, the more accurate the estimate is. If a point estimator is consistent, its expected value and variance should be close to the true value of the parameter. An unbiased estimator is consistent if the limit of the variance of estimator T equals zero.

Efficiency

Let T_1 and T_2 be two unbiased estimators for the same parameter θ. The estimator T_2 would be called *more efficient* than estimator T_1 if $Var(T_2) < Var(T_1)$, irrespective of the value of θ. We can also say that the most efficient estimators are the ones with the least variability of outcomes. Therefore, if the estimator has smallest variance among sample to sample, it is both most efficient and unbiased. We extend the notion of efficiency by saying that estimator T_2 is more efficient than estimator T_1 (for the same parameter of interest), if the MSE(mean square error) of T_2 is smaller than the MSE of T_1.

Generally, we must consider the distribution of the population when determining the efficiency of estimators. For example, in a normal distribution, the mean is considered more efficient than the median, but the same does not apply in asymmetrical, or skewed, distributions.

Sufficiency

In statistics, the job of a statistician is to interpret the data that they have collected and to draw statistically valid conclusion about the population under investigation. But in many cases the raw data, which are too numerous and too costly to store, are not suitable for this purpose. Therefore, the statistician would like to condense the data by computing some statistics and to base their analysis on these statistics so that there is no loss of relevant information in doing so, that is the statistician would like to choose those statistics which exhaust all information about the parameter, which is contained in the sample. We define sufficient statistics as follows: Let X =($X_1, X_2, \dots, X_n$) be a random sample. A statistic T(X) is said to be sufficient for θ (or for the family of distribution) if the conditional distribution of X given T is free from θ.

Types of Point Estimation

Bayesian Point Estimation

Bayesian inference is typically based on the posterior distribution. Many Bayesian point estimators are the posterior distribution's statistics of central tendency, e.g., its mean, median, or mode:

- Posterior mean, which minimizes the (posterior) *risk* (expected loss) for a squared-error loss function; in Bayesian estimation, the risk is defined in terms of the posterior distribution, as observed by Gauss.
- Posterior median, which minimizes the posterior risk for the absolute-value loss function, as observed by Laplace.
- Maximum a posteriori (*MAP*), which finds a maximum of the posterior distribution; for a uniform prior probability, the MAP estimator coincides with the maximum-likelihood estimator.

The MAP estimator has good asymptotic properties, even for many difficult problems, on which the maximum-likelihood estimator has difficulties. For regular problems, where the maximum-likelihood estimator is consistent, the maximum-likelihood estimator ultimately agrees with the MAP estimator. Bayesian estimators are admissible, by Wald's theorem.

The Minimum Message Length (MML) point estimator is based in Bayesian information theory and is not so directly related to the posterior distribution.

Special cases of Bayesian filters are important:

- Kalman filter.
- Wiener filter.

Several methods of computational statistics have close connections with Bayesian analysis:

- Particle filter.
- Markov chain Monte Carlo (MCMC).

Methods of Finding Point Estimates

Below are some commonly used methods of estimating unknown parameters which are expected to provide estimators having some of these important properties. In general, depending on the situation and the purpose of our study we apply any one of the methods that may be suitable among the methods of point estimation.

Method of Maximum Likelihood (MLE)

The method of maximum likelihood, due to R.A. Fisher, is the most important general method of estimation. This estimator method attempts to acquire unknown parameters that maximize the likelihood function. It uses a known model (ex. the normal distribution) and uses the values of parameters in the model that maximize a likelihood function to find the most suitable match for the data.

Let $X = (X_1, X_2, \ldots, X_n)$ denote a random sample with joint p.d.f or p.m.f. $f(x, \theta)$ (θ may be a vector). The function $f(x, \theta)$, considered as a function of θ, is called the likelihood function. In this case, it is denoted by $L(\theta)$. The principle of maximum likelihood consists of choosing an estimate within the admissible range of θ, that maximizes the likelihood. This estimator is called the maximum likelihood estimate (MLE) of θ. In order to obtain the MLE of θ, we use the equation

$d \log L(\theta)/d\theta_i = 0$, $i = 1, 2, \ldots, k$. If θ is a vector, then partial derivatives are considered to get the likelihood equations.

Method of Moments (MOM)

The method of moments was introduced by K. Pearson and P. Chebyshev in 1887, and it is one of the oldest methods of estimation.

This method is based on law of large numbers, which uses all the known facts about a population and apply those facts to a sample of the population by deriving equations that relate the population moments to the unknown parameters. We can then solve with the sample mean of the population moments. However, due to the simplicity, this method is not always accurate and can be biased easily.

Let $(X_1, X_2, \ldots X_n)$ be a random sample from a population having p.d.f. (or p.m.f) $f(x,\theta)$, $\theta = (\theta_1, \theta_2, \ldots, \theta_k)$. The objective is to estimate the parameters $\theta_1, \theta_2, \ldots, \theta_k$. Further, let the first k population moments about zero exist as explicit function of θ, i.e., $\mu_r = \mu_r(\theta_1, \theta_2, \ldots, \theta_k)$, $r = 1, 2, \ldots, k$. In the method of moments, we equate k sample moments with the corresponding population moments.

Generally, the first k moments are taken because the errors due to sampling increase with the order of the moment. Thus, we get k equations $\mu_r(\theta_1, \theta_2, \ldots, \theta_k) =$

m_r, r = 1, 2, ..., k. Solving these equations we get the method of moment estimators (or estimates) as $m_r = 1/n \Sigma X_i^r$.

Method of Least Square

In the method of least square, we consider the estimation of parameters using some specified form of the expectation and second moment of the observations. For fitting a curve of the form $y = f(x, \beta_0, \beta_1, ,,,, \beta_p)$ to the data (x_i, y_i), i = 1, 2,...n, we may use the method of least squares. This method consists of minimizing the sum of squares.

When $f(x, \beta_0, \beta_1, ,,,, \beta_p)$ is a linear function of the parameters and the x-values are known, least square estimators will be best linear unbiased estimator (BLUE). Again, if we assume that the least square estimates are independently and identically normally distributed, then a linear estimator will be minimum-variance unbiased estimator (MVUE) for the entire class of unbiased estimators.

Minimum-Variance Mean-Unbiased Estimator (MVUE)

The method of minimum-variance unbiased estimator minimizes the risk (expected loss) of the squared-error loss-function.

Median Unbiased Estimator

Median-unbiased estimator minimizes the risk of the absolute-error loss function.

Best Linear Unbiased Estimator (BLUE)

Best linear unbiased estimator, also known as the Gauss–Markov theorem states that the ordinary least squares (OLS) estimator has the lowest sampling variance within the class of linear unbiased estimators, if the errors in the linear regression model are uncorrelated, have equal variances and expectation value of zero.

Point Estimate vs. Confidence Interval Estimate

There are two major types of estimates: point estimate and confidence interval estimate. In the point estimate we try to choose a unique point in the parameter space which can reasonably be considered as the true value of the parameter. On the other hand, instead of unique estimate of the parameter, we are interested in constructing a family of sets that contain the true (unknown) parameter value with a specified probability. In many problems of statistical inference we are not interested only in estimating the parameter or testing some hypothesis concerning the parameter, we also want to get a lower or an upper bound or both, for the real-valued parameter. To do this, we need to construct a confidence interval.

Confidence interval describes how reliable an estimate is. We can calculate the upper and lower confidence limits of the intervals from the observed data. Suppose a dataset $x_1, \ldots, x_n$ is given, modeled as realization of random variables $X_1, \ldots, X_n$.

Let θ be the parameter of interest, and γ a number between 0 and 1. If there exist sample statistics $L_n = g(X_1, \ldots, X_n)$ and $U_n = h(X_1, \ldots, X_n)$ such that $P(L_n < \theta < U_n) = \gamma$ for every value of θ, then (l_n, u_n), where $l_n = g(x_1, \ldots, x_n)$ and $u_n = h(x_1, \ldots, x_n)$, is called a $100\gamma\%$ confidence interval for θ. The number γ is called the confidence level. In general, with a normally-distributed sample mean, $\dot{X}$, and with a known value for the standard deviation, σ, a $100(1-\alpha)\%$ confidence interval for the true μ is formed by taking $\dot{X} \pm e$, with $e = z_{1-\alpha/2}(\sigma/n^{1/2})$, where $z_{1-\alpha/2}$ is the $100(1-\alpha/2)\%$ cumulative value of the standard normal curve, and n is the number of data values in that column. For example, $z_{1-\alpha/2}$ equals 1.96 for 95% confidence.

Here two limits are computed from the set of observations, say l_n and u_n and it is claimed with a certain degree of confidence (measured in probabilistic terms) that the true value of γ lies between l_n and u_n. Thus we get an interval (l_n and u_n) which we expect would include the true value of $\gamma(\theta)$. So this type of estimation is called confidence interval estimation. This estimation provides a range of values which the parameter is expected to lie. It generally gives more information than point estimates and are preferred when making inferences. In some way, we can say that point estimation is the opposite of interval estimation.

11.3.2 Confidence Intervals

Informally, in frequentist statistics, a confidence interval (CI) is an interval which is expected to typically contain the parameter being estimated. More specifically, given a confidence level γ (95% and 99% are typical values), a CI is a *random* interval which contains the parameter being estimated γ of the time.

The confidence level, degree of confidence or confidence coefficient represents the long-run proportion of CIs (at the given confidence level) that theoretically contain the true value of the parameter; this is tantamount to the nominal coverage probability. For example, out of all intervals computed at the 95% level, 95% of them should contain the parameter's true value.

Factors affecting the width of the CI include the sample size, the variability in the sample, and the confidence level.

All else being the same, a larger sample produces a narrower confidence interval, greater variability in the sample produces a wider confidence interval, and a higher confidence level produces a wider confidence interval.

History

Methods for calculating confidence intervals for the binomial proportion appeared from the 1920s. The main ideas of confidence intervals in general were developed in the early 1930s, and the first thorough and general account was given by Jerzy Neyman in 1937.

Neyman described the development of the ideas as follows (reference numbers have been changed):

- [My work on confidence intervals] originated about 1930 from a simple question of Waclaw Pytkowski, then my student in Warsaw, engaged in an empirical study in farm economics. The question was: how to characterize non-dogmatically the precision of an estimated regression coefficient? ...
- Pytkowski's monograph ... appeared in print in 1932. It so happened that, somewhat earlier, Fisher published his first paper concerned with fiducial distributions and fiducial argument. Quite unexpectedly, while the conceptual framework of fiducial argument is entirely different from that of confidence intervals, the specific solutions of several particular problems coincided. Thus, in the first paper in which I presented the theory of confidence intervals, published in 1934, I recognized Fisher's priority for the idea that interval estimation is possible without any reference to Bayes' theorem and with the solution being independent from probabilities *a priori*. At the same time I mildly suggested that Fisher's approach to the problem involved a minor misunderstanding.

Desired Properties

When applying standard statistical procedures, there will often be standard ways of constructing confidence intervals. These will have been devised so as to meet certain desirable properties, which will hold given that the assumptions on which the procedure relies are true. These desirable properties may be described as: validity, optimality, and invariance.

Of the three, "validity" is most important, followed closely by "optimality". "Invariance" may be considered as a property of the method of derivation of a confidence interval, rather than of the rule for constructing the interval. In non-standard applications, these same desirable properties would be sought:

Validity

This means that the nominal coverage probability (confidence level) of the confidence interval should hold, either exactly or to a good approximation.

Optimality

This means that the rule for constructing the confidence interval should make as much use of the information in the data-set as possible. One way of assessing optimality is by the width of the interval so that a rule for constructing a confidence interval is judged better than another if it leads to intervals whose widths are typically shorter.

Invariance

In many applications, the quantity being estimated might not be tightly defined as such. For example, a survey might result in an estimate of the median income in a population, but it might equally be considered as providing an estimate of the logarithm of the median income, given that this is a common scale for presenting graphical results.

It would be desirable that the method used for constructing a confidence interval for the median income would give equivalent results when applied to constructing a confidence interval for the logarithm of the median income: Specifically the values at the ends of the latter interval would be the logarithms of the values at the ends of former interval.

Methods of Derivation

For non-standard applications, there are several routes that might be taken to derive a rule for the construction of confidence intervals. Established rules for standard procedures might be justified or explained via several of these routes. Typically a rule for constructing confidence intervals is closely tied to a particular way of finding a point estimate of the quantity being considered.

Summary Statistics

This is closely related to the method of moments for estimation. A simple example arises where the quantity to be estimated is the population mean, in which case a natural estimate is the sample mean. Similarly, the sample variance can be used to estimate the population variance. A confidence interval for the true mean can be constructed centered on the sample mean with a width which is a multiple of the square root of the sample variance.

Likelihood Theory

Estimates can be constructed using the maximum likelihood principle, the likelihood theory for this provides two ways of constructing confidence intervals or confidence regions for the estimates.

Estimating Equations

The estimation approach here can be considered as both a generalization of the method of moments and a generalization of the maximum likelihood approach. There are corresponding generalizations of the results of maximum likelihood theory that allow confidence intervals to be constructed based on estimates derived from estimating equations.

Hypothesis Testing

If hypothesis tests are available for general values of a parameter, then confidence intervals/regions can be constructed by including in the 100 p % confidence region all those points for which the hypothesis test of the null hypothesis that the true value is the given value is not rejected at a significance level of (1 - p).

Bootstrapping

In situations where the distributional assumptions for the above methods are uncertain or violated, resampling methods allow construction of confidence intervals or prediction intervals. The observed data distribution and the internal correlations are used as the surrogate for the correlations in the wider population.

Interpretation

Various interpretations of a confidence interval can be given (taking the 95% confidence interval as an example in the following).

- The confidence interval can be expressed in terms of a long-run frequency in repeated samples (or in resampling): "Were this procedure to be repeated on numerous samples, the proportion of calculated 95% confidence intervals that encompassed the true value of the population parameter would tend toward 95%."
- The confidence interval can be expressed in terms of probability with respect to a single theoretical (yet to be realized) sample: "There is a 95% probability that the 95% confidence interval calculated from a given future sample will cover the true value of the population parameter." This essentially reframes the "repeated samples" interpretation as a probability rather than a frequency.
- The confidence interval can be expressed in terms of statistical significance, e.g.: "The 95% confidence interval represents values that are not statistically significantly different from the point estimate at the .05 level."

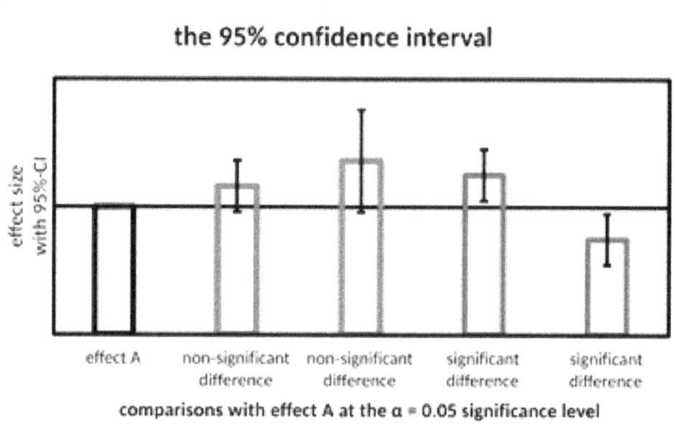

Fig 1: Interpretation of the 95% Confidence Interval in Terms of Statistical Significance

Common Misunderstandings

Confidence intervals and levels are frequently misunderstood, and published studies have shown that even professional scientists often misinterpret them.

- A 95% confidence level does not mean that for a given realized interval there is a 95% probability that the population parameter lies within the interval (i.e., a 95% probability that the interval covers the population parameter). According to the frequentist interpretation, once an interval is calculated, this interval either covers the parameter value or it does not; it is no longer a matter of probability. The 95% probability relates to the reliability of the estimation procedure, not to a specific calculated interval. Neyman himself (the original proponent of confidence intervals) made this point in his original paper:
    - It will be noticed that in the above description, the probability statements refer to the problems of estimation with which the statistician will be concerned in the future. In fact, I have repeatedly stated that the frequency of correct results will tend to α. Consider now the case when a sample is already drawn, and the calculations have given [particular limits]. Can we say that in this particular case the probability of the true value [falling between these limits] is equal to α? The answer is obviously in the negative. The parameter is an unknown constant, and no probability statement concerning its value may be made...
- A 95% confidence level does not mean that 95% of the sample data lie within the confidence interval.
- A 95% confidence level does not mean that there is a 95% probability of the parameter estimate from a repeat of the experiment falling within the confidence interval computed from a given experiment.

11.3.3 Hypothesis Testing

Hypothesis testing is another core component of statistical inference. It is a formal procedure used to test whether the evidence in the sample data supports a specific hypothesis about a population parameter.

The two hypotheses involved are:

- **Null Hypothesis (H_0)**: The statement that there is no effect or no difference. It is the assumption that the observed result is due to random chance.
- **Alternative Hypothesis (H_1 or H_a)**: The statement that there is an effect or a difference. This is what the researcher seeks to support.

In hypothesis testing, a test statistic is calculated, and a p-value is determined. The p-value represents the probability of observing the sample data, or something more extreme, if the null hypothesis is true. If the p-value is less than a predetermined significance level (usually 0.05), the null hypothesis is rejected in favor of the alternative hypothesis.

11.3.4 Types of Errors in Hypothesis Testing

There are two types of errors that can occur in hypothesis testing:

- **Type I Error**: Rejecting the null hypothesis when it is true (a false positive). The probability of making a Type I error is represented by the significance level, α.
- **Type II Error**: Failing to reject the null hypothesis when it is false (a false negative). The probability of making a Type II error is represented by β.

Balancing the risk of Type I and Type II errors is a critical part of designing a hypothesis test.

Researchers typically aim to minimize both errors, but this can involve trade-offs depending on the context of the study.

11.3.5 p-Values

The p-value is a key concept in hypothesis testing and is used to assess the strength of the evidence against the null hypothesis. A smaller p-value indicates stronger evidence that the observed data is not due to chance. In many fields, a p-value of less than 0.05 is considered statistically significant, meaning that the null hypothesis can be rejected. However, it is important to interpret p-values in context, as they do not measure the size of an effect or the importance of a result.

11.4 THE ROLE OF SAMPLE SIZE

Sample size plays a crucial role in statistical inference. Larger samples provide more information and lead to more precise estimates of population parameters. This is because larger samples tend to have less variability and are more representative of the population.

In hypothesis testing, larger sample sizes increase the power of the test, which is the probability of correctly rejecting the null hypothesis when it is false. A larger sample size reduces the risk of a Type II error, making it easier to detect real differences or effects. However, collecting larger samples can be expensive or impractical in some cases, so researchers must balance the need for precision with practical constraints.

11.5 CONCLUSION

Statistical inference is an essential tool for drawing conclusions from data and making informed decisions based on sample information. Through methods such as point estimation, confidence intervals, and hypothesis testing, researchers can make valid inferences about population parameters, even when only a small portion of the population is observed.

By understanding the assumptions and limitations of statistical inference, as well as key concepts like p-values, confidence intervals, and hypothesis testing, analysts can apply these techniques effectively in various fields. Statistical inference, therefore, provides the foundation for evidence-based decision-making in medicine, business, social sciences, and many other disciplines.

CHAPTER 12

TESTS OF SIGNIFICANCE AND ESTIMATION

Tests of significance and estimation are foundational concepts in statistical analysis, used to draw conclusions about populations based on sample data. These concepts are integral to research, particularly in fields such as medicine, social sciences, economics, and engineering, where understanding the likelihood of observed results and the parameters that describe them is essential. Estimation involves calculating the approximate values of population parameters using sample data, while tests of significance assess the strength of evidence against a null hypothesis.

12.1 ESTIMATION

Estimation (or estimating) is the process of finding an estimate or approximation, which is a value that is usable for some purpose even if input data may be incomplete, uncertain, or unstable. The value is nonetheless usable because it is derived from the best information available. Typically, estimation involves "using the value of a statistic derived from a sample to estimate the value of a corresponding population parameter". The sample provides information that can be projected, through various formal or informal processes, to determine a range most likely to describe the missing information. An estimate that turns out to be incorrect will be an overestimate if the estimate exceeds the actual result and an underestimate if the estimate falls short of the actual result.

The confidence in an estimate is quantified as a confidence interval, the likelihood that the estimate is in a certain range. Human estimators systematically suffer from overconfidence, believing that their estimates are more accurate than they actually are.

Estimation is often done by sampling, which is counting a small number of examples something, and projecting that number onto a larger population. An example of estimation would be determining how many candies of a given size are in a glass jar.

Because the distribution of candies inside the jar may vary, the observer can count the number of candies visible through the glass, consider the size of the jar, and presume that a similar distribution can be found in the parts that can not be seen, thereby making an estimate of the total number of candies that could be in the jar if that presumption were true. Estimates can similarly be generated by projecting results from polls or surveys onto the entire population.

In making an estimate, the goal is often most useful to generate a range of possible outcomes that is precise enough to be useful but not so precise that it is likely to be inaccurate. For example, in trying to guess the number of candies in the jar, if fifty were visible, and the total volume of the jar seemed to be about twenty times as large as the volume containing the visible candies, then one might simply project that there were a thousand candies in the jar. Such a projection, intended to pick the single value that is believed to be closest to the actual value, is called a point estimate. However, a point estimation is likely to be incorrect, because the sample size—in this case, the number of candies that are visible—is too small a number to be sure that it does not contain anomalies that differ from the population as a whole. A corresponding concept is an interval estimate, which captures a much larger range of possibilities, but is too broad to be useful. For example, if one were asked to estimate the percentage of people who like candy, it would clearly be correct that the number falls between zero and one hundred percent. Such an estimate would provide no guidance, however, to somebody who is trying to determine how many candies to buy for a party to be attended by a hundred people.

12.1.1 Uses of Estimation

In mathematics, approximation describes the process of finding estimates in the form of upper or lower bounds for a quantity that cannot readily be evaluated precisely, and approximation theory deals with finding simpler functions that are close to some complicated function and that can provide useful estimates. In statistics, an estimator is the formal name for the rule by which an estimate is calculated from data, and estimation theory deals with finding estimates with good properties. This process is used in signal processing, for approximating an unobserved signal on the basis of an observed signal containing noise. For estimation of yet-to-be observed quantities, forecasting and prediction are applied. A Fermi problem, in physics, is one concerning estimation in problems that typically involve making justified guesses about quantities that seem impossible to compute given limited available information.

Estimation is important in business and economics because too many variables exist to figure out how large-scale activities will develop. Estimation in project planning can be particularly significant, because plans for the distribution of labor and purchases of raw materials must be made, despite the inability to know every possible problem that may come up. A certain amount of resources will be available for carrying out a particular project, making it important to obtain or generate a cost estimate as one of the vital elements of entering into the project. The U.S. Government Accountability Office defines a cost estimate as, "the summation of individual cost elements, using established methods and valid data, to estimate the future costs of a program, based on what is known today", and reports that "realistic cost estimating was imperative when making wise decisions in acquiring new systems".

Furthermore, project plans must not underestimate the needs of the project, which can result in delays while unmet needs are fulfilled, nor must they greatly overestimate the needs of the project, or else the unneeded resources may go to waste. An informal estimate when little information is available is called a guesstimate because the inquiry becomes closer to purely guessing the answer. The "estimated" sign, e, is used to designate that package contents are close to the nominal contents.

12.1.2 Point Estimation

Point estimation refers to the use of a single value, derived from a sample, to estimate a population parameter. The most common point estimators are the sample mean (used to estimate the population mean) and the sample proportion (used to estimate the population proportion). For example, in a study of patients' blood pressure, the sample mean blood pressure could be used to estimate the average blood pressure of the entire population.

However, point estimates alone do not convey the uncertainty or variability associated with the estimation. Although they provide the best guess of the population parameter, they lack information about how precise the estimate is.

12.1.3 Interval Estimation

In statistics, interval estimation is the use of sample data to estimate an *interval* of possible values of a parameter of interest. This is in contrast to point estimation, which gives a single value.

The most prevalent forms of interval estimation are *confidence intervals* (a frequentist method) and *credible intervals* (a Bayesian method). Less common forms include *likelihood intervals, fiducial intervals, tolerance intervals,* and *prediction intervals*. For a non-statistical method, interval estimates can be deduced from fuzzy logic.

Types of Interval Estimation

Confidence Intervals

Confidence intervals are used to estimate the parameter of interest from a sampled data set, commonly the mean or standard deviation. A confidence interval states there is a $100\gamma\%$ confidence that the parameter of interest is within a lower and upper bound. A common misconception of confidence intervals is $100\gamma\%$ of the data set fits within or above/below the bounds, this is referred to as a tolerance interval.

There are multiple methods used to build a confidence interval, the correct choice depends on the data being analyzed. For a normal distribution with a known variance, one uses the z-table to create an interval where a confidence level of 100γ% can be obtained centered around the sample mean from a data set of n measurements, . For a Binomial distribution, confidence intervals can be approximated using the Wald Approximate Method, Jeffreys interval, and Clopper-Pearson interval. The Jeffrey method can also be used to approximate intervals for a Poisson distribution. If the underlying distribution is unknown, one can utilize bootstrapping to create bounds about the median of the data set.

Credible Intervals

As opposed to a confidence interval, a credible interval requires a prior assumption, modifying the assumption utilizing a Bayes factor, and determining a posterior distribution. Utilizing the posterior distribution, one can determine a 100γ% *probability* the parameter of interest is included, as opposed to the confidence interval where one can be 100γ% *confident* that an estimate is included within an interval.

$$\text{Posterior} \propto \text{Likelihood} \times \text{Prior}$$

While a prior assumption is helpful towards providing more data towards building an interval, it removes the objectivity of a confidence interval. A prior will be used to inform a posterior, if unchallenged this prior can lead to incorrect predictions.

The credible interval's bounds are variable, unlike the confidence interval. There are multiple methods to determine where the correct upper and lower limits should be located. Common techniques to adjust the bounds of the interval include highest posterior density interval (HPDI), equal-tailed interval, or choosing the center the interval around the mean.

Less Common Forms

- **Likelihood-Based:** Utilizes the principles of a likelihood function to estimate the parameter of interest. Utilizing the likelihood-based method, confidence intervals can be found for exponential, Weibull, and lognormal means. Additionally, likelihood-based approaches can give confidence intervals for the standard deviation. It is also possible to create a prediction interval by combining the likelihood function and the future random variable.
- **Fiducial:** Fiducial inference utilizes a data set, carefully removes the noise and recovers a distribution estimator, Generalized Fiducial Distribution (GFD). Without the use of Bayes' Theorem, there is no assumption of a prior, much like confidence intervals. Fiducial inference is a less common form of statistical inference. The founder, R.A. Fisher,

who had been developing inverse probability methods, had his own questions about the validity of the process. While fiducial inference was developed in the early twentieth century, the late twentieth century believed that the method was inferior to the frequentist and Bayesian approaches but held an important place in historical context for statistical inference. However, modern-day approaches have generalized the fiducial interval into Generalized Fiducial Inference (GFI), which can be used to estimate discrete and continuous data sets.

- **Prediction:** A prediction interval estimates the interval containing future samples with some confidence, γ. Prediction intervals can be used for both Bayesian and frequentist contexts. These intervals are typically used in regression data sets, but prediction intervals are not used for extrapolation beyond the previous data's experimentally controlled parameters.
- **Fuzzy Logic:** Fuzzy logic is used to handle decision-making in a non-binary fashion for artificial intelligence, medical decisions, and other fields. In general, it takes inputs, maps them through fuzzy inference systems, and produces an output decision. This process involves fuzzification, fuzzy logic rule evaluation, and defuzzification. When looking at fuzzy logic rule evaluation, membership functions convert our non-binary input information into tangible variables. These membership functions are essential to predict the uncertainty of the system.

One-Sided vs. Two-Sided

Two-sided intervals estimate a parameter of interest, Θ, with a level of confidence, γ, using a lower and upper bound. Examples may include estimating the average height of males in a geographic region or lengths of a particular desk made by a manufacturer.

These cases tend to estimate the central value of a parameter. Typically, this is presented in a form similar to the equation below.

$$P(l_b < \Theta < u_b) = \gamma$$

Differentiating from the two-sided interval, the one-sided interval utilizes a level of confidence, γ, to construct a minimum or maximum bound which predicts the parameter of interest to γ*100% probability. Typically, a one-sided interval is required when the estimate's minimum or maximum bound is not of interest. When concerned about the minimum predicted value of Θ, one is no longer required to find an upper bounds of the estimate, leading to a form reduced form of the two-sided.

$$P(l_b < \Theta) = \gamma$$

As a result of removing the upper bound and maintaining the confidence, the lower-bound will increase. Likewise, when concerned with finding only an upper bound of a parameter's estimate, the upper bound will decrease. A one-sided interval is a commonly found in material production's quality assurance, where an expected value of a material's strength, Θ, must be above a certain minimum value with some confidence (100γ%). In this case, the manufacturer is not concerned with producing a product that is too strong, there is no upper-bound.

Caution Using and Building Estimates

When determining the significance of a parameter, it is best to understand the data and its collection methods. Before collecting data, an experiment should be planned such that the uncertainty of the data is sample variability, as opposed to a statistical bias. After experimenting, a typical first step in creating interval estimates is plotting using various graphical methods. From this, one can determine the distribution of samples from the data set. Producing interval boundaries with incorrect assumptions based on distribution makes a prediction faulty.

When interval estimates are reported, they should have a commonly held interpretation within and beyond the scientific community. Interval estimates derived from fuzzy logic have much more application-specific meanings.

In commonly occurring situations there should be sets of standard procedures that can be used, subject to the checking and validity of any required assumptions. This applies for both confidence intervals and credible intervals. However, in more novel situations there should be guidance on how interval estimates can be formulated.

In this regard confidence intervals and credible intervals have a similar standing but there two differences. First, credible intervals can readily deal with prior information, while confidence intervals cannot. Secondly, confidence intervals are more flexible and can be used practically in more situations than credible intervals: one area where credible intervals suffer in comparison is in dealing with non-parametric models.

There should be ways of testing the performance of interval estimation procedures. This arises because many such procedures involve approximations of various kinds and there is a need to check that the actual performance of a procedure is close to what is claimed. The use of stochastic simulations makes this is straightforward in the case of confidence intervals, but it is somewhat more problematic for credible intervals where prior information needs to be taken properly into account. Checking of credible intervals can be done for situations representing no-prior-information but the check involves checking the long-run frequency properties of the procedures.

Severini discusses conditions under which credible intervals and confidence intervals will produce similar results, and also discusses both the coverage probabilities of credible intervals and the posterior probabilities associated with confidence intervals.

In decision theory, which is a common approach to and justification for Bayesian statistics, interval estimation is not of direct interest. The outcome is a decision, not an interval estimate, and thus Bayesian decision theorists use a Bayes action: they minimize expected loss of a loss function with respect to the entire posterior distribution, not a specific interval.

Applications

Applications of confidence intervals are used to solve a variety of problems dealing with uncertainty. Katz (1975) proposes various challenges and benefits for utilizing interval estimates in legal proceedings. For use in medical research, Altmen (1990) discusses the use of confidence intervals and guidelines towards using them. In manufacturing, it is also common to find interval estimates estimating a product life, or to evaluate the tolerances of a product. Meeker and Escobar (1998) present methods to analyze reliability data under parametric and nonparametric estimation, including the prediction of future, random variables (prediction intervals).

12.2 TESTS OF SIGNIFICANCE

Tests of significance are statistical procedures used to assess whether the results of an analysis are likely to be due to random chance or whether they provide strong evidence against a null hypothesis. These tests help determine whether observed differences, relationships, or effects in the sample data are statistically significant and can be generalized to the population.

12.2.1 Null and Alternative Hypotheses

Null Hypothesis

In scientific research, the null hypothesis (often denoted H_0) is the claim that the effect being studied does not exist. The null hypothesis can also be described as the hypothesis in which no relationship exists between two sets of data or variables being analyzed. If the null hypothesis is true, any experimentally observed effect is due to chance alone, hence the term "null". In contrast with the null hypothesis, an alternative hypothesis is developed, which claims that a relationship does exist between two variables.

History of Statistical Tests

The history of the null and alternative hypotheses has much to do with the history of statistical tests.

- **Before 1925:** There are occasional transient traces of statistical tests in past centuries, which were early examples of null hypotheses. In the late 19th century statistical significance was defined. In the early 20th century important probability distributions were defined. Gossett and Pearson worked on specific cases of significance testing.
- **1925:** Fisher published the first edition of *Statistical Methods for Research Workers,* which defined the statistical significance test and made it a mainstream method of analysis for much of experimental science. The text was devoid of proofs and weak on explanations, but was filled with real examples. It placed statistical practice in the sciences well in advance of published statistical theory.
- **1933:** In a series of papers (published over a decade starting in 1928) Neyman & Pearson defined the statistical hypothesis test as a proposed improvement on Fisher's test. The papers provided much of the terminology for statistical tests including *alternative hypothesis* and H_0 as a hypothesis to be tested using observational data (with H_1, H_2... as alternatives).
- **1935:** Fisher published the first edition of the book *The Design of Experiments* which introduced the null hypothesis (by example rather than by definition) and carefully explained the rationale for significance tests in the context of the interpretation of experimental results.
- Fisher and Neyman quarreled over the relative merits of their competing formulations until Fisher's death in 1962. Career changes and World War II ended the partnership of Neyman and Pearson. The formulations were merged by relatively anonymous textbook writers, experimenters (journal editors) and mathematical statisticians without input from either Fisher or Neyman. The subject today combines much of the terminology and explanatory power of Neyman & Pearson with the scientific philosophy and calculations provided by Fisher. Whether statistical testing is properly one subject or two remains a source of disagreement. Sample of two: One text refers to the subject as hypothesis testing (with no mention of significance testing in the index) while another says significance testing (with a section on inference as a decision). Fisher developed significance testing as a flexible tool for researchers to weigh their evidence. Instead testing has become institutionalized. Statistical significance has become a rigidly defined and enforced criterion for the publication of experimental results in many scientific journals. In some fields significance testing has become the dominant and nearly exclusive form of statistical analysis. As a

consequence the limitations of the tests have been exhaustively studied. Books have been filled with the collected criticism of significance testing.

Basic Definitions

The null hypothesis and the *alternative hypothesis* are types of conjectures used in statistical tests to make statistical inferences, which are formal methods of reaching conclusions and separating scientific claims from statistical noise.

The statement being tested in a test of statistical significance is called the null hypothesis. The test of significance is designed to assess the strength of the evidence against the null hypothesis, or a statement of 'no effect' or 'no difference'. It is often symbolized as H_0. The statement that is being tested against the null hypothesis is the alternative hypothesis. Symbols may include H_1 and H_a.

A statistical significance test starts with a random sample from a population. If the sample data are consistent with the null hypothesis, then you do not reject the null hypothesis; if the sample data are inconsistent with the null hypothesis, then you reject the null hypothesis and conclude that the alternative hypothesis is true.

Consider the following example. Given the test scores of two random samples, one of men and one of women, does one group score better than the other?

A possible null hypothesis is that the mean male score is the same as the mean female score:

$$H_0: \mu_1 = \mu_2$$

where

- H_0 = the null hypothesis,
- μ_1 = the mean of population 1, and
- μ_2 = the mean of population 2.

A stronger null hypothesis is that the two samples have equal variances and shapes of their respective distributions.

The simple/composite distinction was made by Neyman and Pearson.

- **Exact Hypothesis:** Any hypothesis that specifies an exact parameter value. Example: $\mu = 100$. Synonym: point hypothesis.
- **Inexact Hypothesis:** Those specifying a parameter range or interval. Examples: $\mu \leq 100$; $95 \leq \mu \leq 105$.

Fisher required an exact null hypothesis for testing. A one-tailed hypothesis (tested using a one-sided test) is an inexact hypothesis in which the value of a parameter is specified as being either:

- Above or equal to a certain value, or
- Below or equal to a certain value.

A one-tailed hypothesis is said to have directionality. Fisher's original (lady tasting tea) example was a one-tailed test. The null hypothesis was asymmetric. The probability of guessing all cups correctly was the same as guessing all cups incorrectly, but Fisher noted that only guessing correctly was compatible with the lady's claim.

Technical Description

The null hypothesis is a default hypothesis that a quantity to be measured is zero (null). Typically, the quantity to be measured is the difference between two situations. For instance, trying to determine if there is a positive proof that an effect has occurred or that samples derive from different batches.

The null hypothesis is generally assumed to remain possibly true. Multiple analyses can be performed to show how the hypothesis should either be rejected or excluded e.g. having a high confidence level, thus demonstrating a statistically significant difference. This is demonstrated by showing that zero is outside of the specified confidence interval of the measurement on either side, typically within the real numbers. Failure to exclude the null hypothesis (with any confidence) does not logically confirm or support the (unprovable) null hypothesis. (When it is proven that something is e.g. bigger than x, it does not necessarily imply it is plausible that it is smaller or equal than x; it may instead be a poor quality measurement with low accuracy.

Confirming the null hypothesis two-sided would amount to positively proving it is bigger or equal than 0 *and* to positively proving it is smaller or equal than 0; this is something for which infinite accuracy is needed as well as exactly zero effect, neither of which normally are realistic. Also measurements will never indicate a non-zero probability of exactly zero difference.) So failure of an exclusion of a null hypothesis amounts to a "don't know" at the specified confidence level; it does not immediately imply null somehow, as the data may already show a (less strong) indication for a non-null. The used confidence level does absolutely certainly not correspond to the likelihood of null at failing to exclude; in fact in this case a high used confidence level *expands* the still plausible range.

A non-null hypothesis can have the following meanings, depending on the author a) a value other than zero is used, b) some margin other than zero is used and c) the "alternative" hypothesis.

Testing (excluding or failing to exclude) the null hypothesis provides evidence that there are (or are not) statistically sufficient grounds to believe there *is* a relationship between two phenomena (e.g., that a potential treatment has a non-zero effect, either way). Testing the null hypothesis is a central task in statistical hypothesis testing in the modern practice of science. There are precise criteria for excluding or not excluding a null hypothesis at a certain confidence level. The confidence level should indicate the likelihood that much more and better data would still be able to exclude the null hypothesis on the same side.

The concept of a null hypothesis is used differently in two approaches to statistical inference. In the significance testing approach of Ronald Fisher, a null hypothesis is rejected if the observed data are significantly unlikely to have occurred if the null hypothesis were true. In this case, the null hypothesis is rejected and an alternative hypothesis is accepted in its place. If the data are consistent with the null hypothesis statistically possibly true, then the null hypothesis is not rejected. In neither case is the null hypothesis or its alternative proven; with better or more data, the null may still be rejected. This is analogous to the legal principle of presumption of innocence, in which a suspect or defendant is assumed to be innocent (null is not rejected) until proven guilty (null is rejected) beyond a reasonable doubt (to a statistically significant degree).

In the hypothesis testing approach of Jerzy Neyman and Egon Pearson, a null hypothesis is contrasted with an alternative hypothesis, and the two hypotheses are distinguished on the basis of data, with certain error rates. It is used in formulating answers in research.

Statistical inference can be done without a null hypothesis, by specifying a statistical model corresponding to each candidate hypothesis, and by using model selection techniques to choose the most appropriate model. (The most common selection techniques are based on either Akaike information criterion or Bayes factor).

Principle

Hypothesis testing requires constructing a statistical model of what the data would look like if chance or random processes alone were responsible for the results. The hypothesis that chance alone is responsible for the results is called the *null hypothesis*. The model of the result of the random process is called the *distribution under the null hypothesis*. The obtained results are compared with the distribution under the null hypothesis, and the likelihood of finding the obtained results is thereby determined.

Hypothesis testing works by collecting data and measuring how likely the particular set of data is (assuming the null hypothesis is true), when the study is on a randomly selected representative sample. The null hypothesis assumes no relationship between variables in the population from which the sample is selected.

If the data-set of a randomly selected representative sample is very unlikely relative to the null hypothesis (defined as being part of a class of sets of data that only rarely will be observed), the experimenter rejects the null hypothesis, concluding it (probably) is false. This class of data-sets is usually specified via a test statistic, which is designed to measure the extent of apparent departure from the null hypothesis. The procedure works by assessing whether the observed departure, measured by the test statistic, is larger than a value defined, so that the probability of occurrence of a more extreme value is small under the null hypothesis (usually in less than either 5% or 1% of similar data-sets in which the null hypothesis does hold).

If the data do not contradict the null hypothesis, then only a weak conclusion can be made: namely, that the observed data set provides insufficient evidence against the null hypothesis. In this case, because the null hypothesis could be true or false, in some contexts this is interpreted as meaning that the data give insufficient evidence to make any conclusion, while in other contexts, it is interpreted as meaning that there is not sufficient evidence to support changing from a currently useful regime to a different one. Nevertheless, if at this point the effect appears likely and/or large enough, there may be an incentive to further investigate, such as running a bigger sample.

For instance, a certain drug may reduce the risk of having a heart attack. Possible null hypotheses are "this drug does not reduce the risk of having a heart attack" or "this drug has no effect on the risk of having a heart attack". The test of the hypothesis consists of administering the drug to half of the people in a study group as a controlled experiment. If the data show a statistically significant change in the people receiving the drug, the null hypothesis is rejected.

Goals of Null Hypothesis Tests

There are many types of significance tests for one, two or more samples, for means, variances and proportions, paired or unpaired data, for different distributions, for large and small samples; all have null hypotheses. There are also at least four goals of null hypotheses for significance tests:

- Technical null hypotheses are used to verify statistical assumptions. For example, the residuals between the data and a statistical model cannot be distinguished from random noise. If true, there is no justification for complicating the model.
- Scientific null assumptions are used to directly advance a theory. For example, the angular momentum of the universe is zero. If not true, the theory of the early universe may need revision.
- Null hypotheses of homogeneity are used to verify that multiple experiments are producing consistent results. For example, the effect of a medication on the elderly is consistent with that of the general

adult population. If true, this strengthens the general effectiveness conclusion and simplifies recommendations for use.

Null hypotheses that assert the equality of effect of two or more alternative treatments, for example, a drug and a placebo, are used to reduce scientific claims based on statistical noise. This is the most popular null hypothesis; It is so popular that many statements about significant testing assume such null hypotheses.

Rejection of the null hypothesis is *not necessarily* the real goal of a significance tester. An adequate statistical model may be associated with a failure to reject the null; the model is adjusted until the null is not rejected. The numerous uses of significance testing were well known to Fisher who discussed many in his book written a decade before defining the null hypothesis.

A statistical significance test shares much mathematics with a confidence interval. They are mutually illuminating. A result is often significant when there is confidence in the sign of a relationship (the interval does not include 0). Whenever the sign of a relationship is important, statistical significance is a worthy goal. This also reveals weaknesses of significance testing: A result can be significant without a good estimate of the strength of a relationship; significance can be a modest goal. A weak relationship can also achieve significance with enough data. Reporting both significance and confidence intervals is commonly recommended. The varied uses of significance tests reduce the number of generalizations that can be made about all applications.

Choice of the Null Hypothesis

The choice of the null hypothesis is associated with sparse and inconsistent advice. Fisher mentioned few constraints on the choice and stated that many null hypotheses should be considered and that many tests are possible for each. The variety of applications and the diversity of goals suggests that the choice can be complicated. In many applications the formulation of the test is traditional. A familiarity with the range of tests available may suggest a particular null hypothesis and test. Formulating the null hypothesis is not automated (though the calculations of significance testing usually are). David Cox said, “How [the] translation from subject-matter problem to statistical model is done is often the most critical part of an analysis”. A statistical significance test is intended to test a hypothesis. If the hypothesis summarizes a set of data, there is no value in testing the hypothesis on that set of data. Example: If a study of last year’s weather reports indicates that rain in a region falls primarily on weekends, it is only valid to test that null hypothesis on weather reports from any *other* year. Testing hypotheses suggested by the data is circular reasoning that proves nothing; It is a special limitation on the choice of the null hypothesis.

A routine procedure is as follows: Start from the scientific hypothesis. Translate this to a statistical alternative hypothesis and proceed: "Because H_a expresses the effect that we wish to find evidence for, we often begin with H_a and then set up H_0 as the statement that the hoped-for effect is not present." This advice is *reversed* for modeling applications where we hope not to find evidence against the null.

A complex case example is as follows: The gold standard in clinical research is the randomized placebo-controlled double-blind clinical trial. But testing a new drug against a (medically ineffective) placebo may be unethical for a serious illness. Testing a new drug against an older medically effective drug raises fundamental philosophical issues regarding the goal of the test and the motivation of the experimenters.

The standard "no difference" null hypothesis may reward the pharmaceutical company for gathering inadequate data. "Difference" is a better null hypothesis in this case, but statistical significance is not an adequate criterion for reaching a nuanced conclusion which requires a good numeric estimate of the drug's effectiveness. A "minor" or "simple" proposed change in the null hypothesis ((new vs old) rather than (new vs placebo)) can have a dramatic effect on the utility of a test for complex non-statistical reasons.

Tailedness of the Null-Hypothesis Test

Consider the question of whether a tossed coin is fair (i.e. that on average it lands heads up 50% of the time) and an experiment where you toss the coin 5 times. A possible result of the experiment that we consider here is 5 heads. Let outcomes be considered unlikely with respect to an assumed distribution if their probability is lower than a significance threshold of 0.05.

A potential null hypothesis implying a one-tailed test is "this coin is not biased toward heads". Beware that, in this context, the term "one-tailed" does *not* refer to the outcome of a single coin toss (i.e., whether or not the coin comes up "tails" instead of "heads"); the term "one-tailed" refers to a specific way of testing the null hypothesis in which the critical region (also known as "region of rejection") ends up in on only one side of the probability distribution.

Indeed, with a fair coin the probability of this experiment outcome is $1/2^5 = 0.031$, which would be even lower if the coin were biased in favour of tails. Therefore, the observations are not likely enough for the null hypothesis to hold, and the test refutes it. Since the coin is ostensibly neither fair nor biased toward tails, the conclusion of the experiment is that the coin is biased towards heads.

Alternatively, a null hypothesis implying a two-tailed test is "this coin is fair". This one null hypothesis could be examined by looking out for either too many tails or too many heads in the experiments. The outcomes that would tend to refute this null hypothesis are those with a large number of heads or a large number of tails, and our experiment with 5 heads would seem to belong to this class.

However, the probability of 5 tosses of the same kind, irrespective of whether these are head or tails, is twice as much as that of the 5-head occurrence singly considered. Hence, under this two-tailed null hypothesis, the observation receives a probability value of 0.063. Hence again, with the same significance threshold used for the one-tailed test (0.05), the same outcome is not statistically significant. Therefore, the two-tailed null hypothesis will be preserved in this case, not supporting the conclusion reached with the single-tailed null hypothesis, that the coin is biased towards heads.

This example illustrates that the conclusion reached from a statistical test may depend on the precise formulation of the null and alternative hypotheses.

Discussion

Fisher said, "the null hypothesis must be exact, that is free of vagueness and ambiguity, because it must supply the basis of the 'problem of distribution,' of which the test of significance is the solution", implying a more restrictive domain for H_0. According to this view, the null hypothesis must be numerically exact—it must state that a particular quantity or difference is equal to a particular number. In classical science, it is most typically the statement that there is *no effect* of a particular treatment; in observations, it is typically that there is *no difference* between the value of a particular measured variable and that of a prediction.

Most statisticians believe that it is valid to state direction as a part of null hypothesis, or as part of a null hypothesis/alternative hypothesis pair. However, the results are not a full description of all the results of an experiment, merely a single result tailored to one particular purpose. For example, consider an H_0 that claims the population mean for a new treatment is an improvement on a well-established treatment with population mean = 10 (known from long experience), with the one-tailed alternative being that the new treatment's mean > 10. If the sample evidence obtained through *x*-bar equals -200 and the corresponding t-test statistic equals -50, the conclusion from the test would be that there is no evidence that the new treatment is better than the existing one: it would not report that it is markedly worse, but that is not what this particular test is looking for. To overcome any possible ambiguity in reporting the result of the test of a null hypothesis, it is best to indicate whether the test was two-sided and, if one-sided, to include the direction of the effect being tested.

The statistical theory required to deal with the simple cases of directionality dealt with here, and more complicated ones, makes use of the concept of an unbiased test.

The directionality of hypotheses is not always obvious. The explicit null hypothesis of Fisher's Lady tasting tea example was that the Lady had no such ability, which led to a symmetric probability distribution. The one-tailed nature of the test resulted from the one-tailed alternate hypothesis (a term not used by Fisher). The null hypothesis became implicitly one-tailed. The logical negation of the Lady's one-tailed claim was also one-tailed. (Claim: Ability > 0; Stated null: Ability = 0; Implicit null: Ability ≤ 0).

Pure arguments over the use of one-tailed tests are complicated by the variety of tests. Some tests (for instance the χ^2 goodness of fit test) are inherently one-tailed. Some probability distributions are asymmetric. The traditional tests of 3 or more groups are two-tailed.

Advice concerning the use of one-tailed hypotheses has been inconsistent and accepted practice varies among fields. The greatest objection to one-tailed hypotheses is their potential subjectivity. A non-significant result can sometimes be converted to a significant result by the use of a one-tailed hypothesis (as the fair coin test, at the whim of the analyst). The flip side of the argument: One-sided tests are less likely to ignore a real effect. One-tailed tests can suppress the publication of data that differs in sign from predictions. Objectivity was a goal of the developers of statistical tests.

It is a common practice to use a one-tailed hypothesis by default. However, "If you do not have a specific direction firmly in mind in advance, use a two-sided alternative. Moreover, some users of statistics argue that we should *always* work with the two-sided alternative."

One alternative to this advice is to use three-outcome tests. It eliminates the issues surrounding directionality of hypotheses by testing twice, once in each direction and combining the results to produce three possible outcomes. Variations on this approach have a history, being suggested perhaps 10 times since 1950.

Disagreements over one-tailed tests flow from the philosophy of science. While Fisher was willing to ignore the unlikely case of the Lady guessing all cups of tea incorrectly (which may have been appropriate for the circumstances), medicine believes that a proposed treatment that kills patients is significant in every sense and should be reported and perhaps explained. Poor statistical reporting practices have contributed to disagreements over one-tailed tests. Statistical significance resulting from two-tailed tests is insensitive to the sign of the relationship; Reporting significance alone is inadequate. "The treatment has an effect" is the uninformative result of a two-tailed test. "The treatment has a beneficial effect" is the more informative result of a one-tailed test. "The treatment has an effect, reducing the average length of hospitalization by 1.5 days" is the most informative report, combining a two-tailed significance test result with a numeric estimate of the relationship between treatment and effect. Explicitly reporting a numeric result eliminates a philosophical advantage of a one-tailed test. An underlying issue is the appropriate form of an experimental science without numeric predictive theories: A model of numeric results is more informative than a model of effect signs (positive, negative or unknown) which is more informative than a model of simple significance (non-zero or unknown); in the absence of numeric theory signs may suffice.

Alternative Hypothesis

In statistical hypothesis testing, the alternative hypothesis is one of the proposed proposition in the hypothesis test. In general the goal of hypothesis test is to demonstrate that in the given condition, there is sufficient evidence supporting the credibility of alternative hypothesis instead of the exclusive proposition in the test (null hypothesis). It is usually consistent with the research hypothesis because it is constructed from literature review, previous studies, etc. However, the research hypothesis is sometimes consistent with the null hypothesis.

In statistics, alternative hypothesis is often denoted as H_a or H_1. Hypotheses are formulated to compare in a statistical hypothesis test. In the domain of inferential statistics, two rival hypotheses can be compared by explanatory power and predictive power.

History

The concept of an alternative hypothesis in testing was devised by Jerzy Neyman and Egon Pearson, and it is used in the Neyman–Pearson lemma. It forms a major component in modern statistical hypothesis testing. However it was not part of Ronald Fisher's formulation of statistical hypothesis testing, and he opposed its use. In Fisher's approach to testing, the central idea is to assess whether the observed dataset could have resulted from chance if the null hypothesis were assumed to hold, notionally without preconceptions about what other models might hold. Modern statistical hypothesis testing accommodates this type of test since the alternative hypothesis can be just the negation of the null hypothesis.

Basic Definition

The *alternative hypothesis* and *null hypothesis* are types of conjectures used in statistical tests, which are formal methods of reaching conclusions or making judgments on the basis of data. In statistical hypothesis testing, the null hypothesis and alternative hypothesis are two mutually exclusive statements.

"The statement being tested in a test of statistical significance is called the null hypothesis. The test of significance is designed to assess the strength of the evidence against the null hypothesis. Usually, the null hypothesis is a statement of 'no effect' or 'no difference'." Null hypothesis is often denoted as H_0. The statement that is being tested against the null hypothesis is the alternative hypothesis. Alternative hypothesis is often denoted as H_a or H_1.

In statistical hypothesis testing, to prove the alternative hypothesis is true, it should be shown that the data is contradictory to the null hypothesis. Namely, there is sufficient evidence against null hypothesis to demonstrate that the alternative hypothesis is true.

Example

One example is where water quality in a stream has been observed over many years, and a test is made of the null hypothesis that "there is no change in quality between the first and second halves of the data", against the alternative hypothesis that "the quality is poorer in the second half of the record".

If the statistical hypothesis testing is thought of as a judgement in a court trial, the null hypothesis corresponds to the position of the defendant (the defendant is innocent) while the alternative hypothesis is in the rival position of prosecutor (the defendant is guilty). The defendant is innocent until proven guilty, so likewise in a hypothesis test, the null hypothesis is initially presumed to be true. To prove the statement of the prosecutor, evidence must be convincing enough to convict the defendant; this is analogous to sufficient statistical significance in a hypothesis test.

In the court, only legal evidence can be considered as the foundation for the trial. As for hypothesis testing, a reasonable test statistic should be set to measure the statistic significance of the null hypothesis. Evidence would support the alternative hypothesis if the null hypothesis is rejected at a certain significance level. However, this does not necessarily mean that the alternative hypothesis is true due to the potential presence of a type I error. In order to quantify the statistical significance, the test statistic variables are assumed to follow a certain probability distribution such as the normal distribution or t-distribution to determine the probability of obtaining test results at least as extreme as the results actually observed, under the assumption that the null hypothesis is correct, which is defined as the *p*-value. If the p-value is smaller than the chosen significance level (α), it can be claimed that observed data is sufficiently inconsistent with the null hypothesis and hence the null hypothesis may be rejected. After testing, a valid claim would be "at the significance level of (α), the null hypothesis is rejected, supporting the alternative hypothesis instead". In the metaphor of a trial, the announcement may be "with tolerance for the probability α of an incorrect conviction, the defendant is guilty."

Types

In the case of a scalar parameter, there are four principal types of alternative hypothesis:

- **Point:** Point alternative hypotheses occur when the hypothesis test is framed so that the population distribution under the alternative hypothesis is a fully defined distribution, with no unknown parameters; such hypotheses are usually of no practical interest but are fundamental to theoretical considerations of statistical inference and are the basis of the Neyman–Pearson lemma.

- **One-Tailed Directional:** A one-tailed directional alternative hypothesis is concerned with the region of rejection for only one tail of the sampling distribution.
- **Two-Tailed Directional:** A two-tailed directional alternative hypothesis is concerned with both regions of rejection of the sampling distribution.
- **Non-Directional:** A non-directional alternative hypothesis is not concerned with either region of rejection; rather, it is only concerned that null hypothesis is not true.

12.2.2 Test Statistic

Test statistic is a quantity derived from the sample for statistical hypothesis testing. A hypothesis test is typically specified in terms of a test statistic, considered as a numerical summary of a data-set that reduces the data to one value that can be used to perform the hypothesis test. In general, a test statistic is selected or defined in such a way as to quantify, within observed data, behaviours that would distinguish the null from the alternative hypothesis, where such an alternative is prescribed, or that would characterize the null hypothesis if there is no explicitly stated alternative hypothesis.

An important property of a test statistic is that its sampling distribution under the null hypothesis must be calculable, either exactly or approximately, which allows p-values to be calculated. A *test statistic* shares some of the same qualities of a descriptive statistic, and many statistics can be used as both test statistics and descriptive statistics. However, a test statistic is specifically intended for use in statistical testing, whereas the main quality of a descriptive statistic is that it is easily interpretable. Some informative descriptive statistics, such as the sample range, do not make good test statistics since it is difficult to determine their sampling distribution. Two widely used test statistics are the t-statistic and the F-statistic.

Example

Suppose the task is to test whether a coin is fair (i.e. has equal probabilities of producing a head or a tail). If the coin is flipped 100 times and the results are recorded, the raw data can be represented as a sequence of 100 heads and tails. If there is interest in the marginal probability of obtaining a tail, only the number T out of the 100 flips that produced a tail needs to be recorded. But T can also be used as a test statistic in one of two ways:

- The exact sampling distribution of T under the null hypothesis is the binomial distribution with parameters 0.5 and 100.
- The value of T can be compared with its expected value under the null hypothesis of 50, and since the sample size is large, a normal distribution can be used as an approximation to the sampling distribution either for T or for the revised test statistic T–50.

Using one of these sampling distributions, it is possible to compute either a one-tailed or two-tailed p-value for the null hypothesis that the coin is fair. The test statistic in this case reduces a set of 100 numbers to a single numerical summary that can be used for testing.

Common Test Statistics

- One-sample tests are appropriate when a sample is being compared to the population from a hypothesis. The population characteristics are known from theory or are calculated from the population.
- Two-sample tests are appropriate for comparing two samples, typically experimental and control samples from a scientifically controlled experiment.
- Paired tests are appropriate for comparing two samples where it is impossible to control important variables. Rather than comparing two sets, members are paired between samples so the difference between the members becomes the sample. Typically the mean of the differences is then compared to zero. The common example scenario for when a paired difference test is appropriate is when a single set of test subjects has something applied to them and the test is intended to check for an effect.

Z-tests are appropriate for comparing means under stringent conditions regarding normality and a known standard deviation. A *t*-test is appropriate for comparing means under relaxed conditions (less is assumed). Tests of proportions are analogous to tests of means (the 50% proportion).

Chi-squared tests use the same calculations and the same probability distribution for different applications:

- Chi-squared tests for variance are used to determine whether a normal population has a specified variance. The null hypothesis is that it does.
- Chi-squared tests of independence are used for deciding whether two variables are associated or are independent. The variables are categorical rather than numeric. It can be used to decide whether left-handedness is correlated with height (or not). The null hypothesis is

that the variables are independent. The numbers used in the calculation are the observed and expected frequencies of occurrence.

- Chi-squared goodness of fit tests are used to determine the adequacy of curves fit to data. The null hypothesis is that the curve fit is adequate. It is common to determine curve shapes to minimize the mean square error, so it is appropriate that the goodness-of-fit calculation sums the squared errors.

F-tests (analysis of variance, ANOVA) are commonly used when deciding whether groupings of data by category are meaningful. If the variance of test scores of the left-handed in a class is much smaller than the variance of the whole class, then it may be useful to study lefties as a group. The null hypothesis is that two variances are the same – so the proposed grouping is not meaningful.

12.2.3 P-Values

In null-hypothesis significance testing, the p p-value is the probability of obtaining test results at least as extreme as the result actually observed, under the assumption that the null hypothesis is correct. A very small *p*-value means that such an extreme observed outcome would be very unlikely under the null hypothesis. Even though reporting *p*-values of statistical tests is common practice in academic publications of many quantitative fields, misinterpretation and misuse of p-values is widespread and has been a major topic in mathematics and metascience. In 2016, the American Statistical Association (ASA) made a formal statement that "*p*-values do not measure the probability that the studied hypothesis is true, or the probability that the data were produced by random chance alone" and that "a *p*-value, or statistical significance, does not measure the size of an effect or the importance of a result" or "evidence regarding a model or hypothesis". That said, a 2019 task force by ASA has issued a statement on statistical significance and replicability, concluding with: "*p*-values and significance tests, when properly applied and interpreted, increase the rigor of the conclusions drawn from data".

Basic Concepts

In statistics, every conjecture concerning the unknown probability distribution of a collection of random variables representing the observed data X X in some study is called a *statistical hypothesis.* If we state one hypothesis only and the aim of the statistical test is to see whether this hypothesis is tenable, but not to investigate other specific hypotheses, then such a test is called a null hypothesis test.

As our statistical hypothesis will, by definition, state some property of the distribution, the null hypothesis is the default hypothesis under which that property does not exist. The null hypothesis is typically that some parameter (such as a correlation or a difference between means) in the populations of interest is zero. Our hypothesis might specify the probability distribution of X X precisely, or it might only specify that it belongs to some class of distributions. Often, we reduce the data to a single numerical statistic, e.g., T T, whose marginal probability distribution is closely connected to a main question of interest in the study.

The *p*-value is used in the context of null hypothesis testing in order to quantify the statistical significance of a result, the result being the observed value of the chosen statistic T T. The lower the *p*-value is, the lower the probability of getting that result if the null hypothesis were true. A result is said to be *statistically significant* if it allows us to reject the null hypothesis. All other things being equal, smaller *p*-values are taken as stronger evidence against the null hypothesis.

Loosely speaking, rejection of the null hypothesis implies that there is sufficient evidence against it.

Usage

The *p*-value is widely used in statistical hypothesis testing, specifically in null hypothesis significance testing. In this method, before conducting the study, one first chooses a model (the null hypothesis) and the alpha level α (most commonly 0.05). After analyzing the data, if the *p*-value is less than α, that is taken to mean that the observed data is sufficiently inconsistent with the null hypothesis for the null hypothesis to be rejected. However, that does not prove that the null hypothesis is false. The *p*-value does not, in itself, establish probabilities of hypotheses. Rather, it is a tool for deciding whether to reject the null hypothesis.

Misuse

According to the ASA, there is widespread agreement that *p*-values are often misused and misinterpreted. One practice that has been particularly criticized is accepting the alternative hypothesis for any *p*-value nominally less than 0.05 without other supporting evidence. Although *p*-values are helpful in assessing how incompatible the data are with a specified statistical model, contextual factors must also be considered, such as "the design of a study, the quality of the measurements, the external evidence for the phenomenon under study, and the validity of assumptions that underlie the data analysis". Another concern is that the *p*-value is often misunderstood as being the probability that the null hypothesis is true.

Some statisticians have proposed abandoning *p*-values and focusing more on other inferential statistics, such as confidence intervals, likelihood ratios, or Bayes factors, but there is heated debate on the feasibility of these alternatives. Others have suggested to remove fixed significance thresholds and to interpret *p*-values as continuous indices of the strength of evidence against the null hypothesis. Yet others suggested to report alongside *p*-values the prior probability of a real effect that would be required to obtain a false positive risk (i.e. the probability that there is no real effect) below a pre-specified threshold (e.g. 5%).

That said, in 2019 a task force by ASA had convened to consider the use of statistical methods in scientific studies, specifically hypothesis tests and *p*-values, and their connection to replicability. It states that "Different measures of uncertainty can complement one another; no single measure serves all purposes", citing *p*-value as one of these measures. They also stress that *p*-values can provide valuable information when considering the specific value as well as when compared to some threshold. In general, it stresses that "*p*-values and significance tests, when properly applied and interpreted, increase the rigor of the conclusions drawn from data".

Calculation

Usually, T T is a test statistic. A test statistic is the output of a scalar function of all the observations. This statistic provides a single number, such as a *t*-statistic or an *F*-statistic. As such, the test statistic follows a distribution determined by the function used to define that test statistic and the distribution of the input observational data.

For the important case in which the data are hypothesized to be a random sample from a normal distribution, depending on the nature of the test statistic and the hypotheses of interest about its distribution, different null hypothesis tests have been developed. Some such tests are the *z*-test for hypotheses concerning the mean of a normal distribution with known variance, the *t*-test based on Student's *t*-distribution of a suitable statistic for hypotheses concerning the mean of a normal distribution when the variance is unknown, the *F*-test based on the *F*-distribution of yet another statistic for hypotheses concerning the variance. For data of other nature, for instance, categorical (discrete) data, test statistics might be constructed whose null hypothesis distribution is based on normal approximations to appropriate statistics obtained by invoking the central limit theorem for large samples, as in the case of Pearson's chi-squared test.

Thus computing a *p*-value requires a null hypothesis, a test statistic (together with deciding whether the researcher is performing a one-tailed test or a two-tailed test), and data. Even though computing the test statistic on given data may be easy, computing the sampling distribution under the null hypothesis, and then computing its cumulative distribution function (CDF) is often a difficult problem. Today, this computation is done using statistical software, often via numeric methods (rather than exact formulae), but, in the early and mid 20th century, this was instead done via tables of values, and one interpolated or extrapolated *p*-values from these discrete values. Rather than using a table of *p*-values, Fisher instead inverted the CDF, publishing a list of values of the test statistic for given fixed *p*-values; this corresponds to computing the quantile function (inverse CDF).

Related Indices

The *E-value* can refer to two concepts, both of which are related to the p-value and both of which play a role in multiple testing. First, it corresponds to a generic, more robust alternative to the p-value that can deal with *optional continuation* of experiments. Second, it is also used to abbreviate "expect value", which is the expected number of times that one expects to obtain a test statistic at least as extreme as the one that was actually observed if one assumes that the null hypothesis is true. This expect-value is the product of the number of tests and the *p*-value. The *q*-value is the analog of the *p*-value with respect to the positive false discovery rate. It is used in multiple hypothesis testing to maintain statistical power while minimizing the false positive rate. The Probability of Direction (*pd*) is the Bayesian numerical equivalent of the *p*-value. It corresponds to the proportion of the posterior distribution that is of the median's sign, typically varying between 50% and 100%, and representing the certainty with which an effect is positive or negative. Second-generation p-values extend the concept of p-values by not considering extremely small, practically irrelevant effect sizes as significant.

12.2.4 Types of Errors in Hypothesis Testing

In statistical hypothesis testing, a type I error, or a false positive, is the rejection of the null hypothesis when it is actually true. For example, an innocent person may be convicted. A type II error, or a false negative, is the failure to reject a null hypothesis that is actually false. For example: a guilty person may be not convicted.

Much of statistical theory revolves around the minimization of one or both of these errors, though the complete elimination of either is an impossibility if the outcome is not determined by a known, observable causal process. By selecting a low threshold (cut-off) value and modifying the alpha (α) level, the quality of the hypothesis test can be increased. The knowledge of type I errors and type II errors is widely used in medical science, biometrics and computer science.

Intuitively, type I errors can be thought of as errors of *commission* (i.e., the researcher unluckily concludes that something is the fact). For instance, consider a study where researchers compare a drug with a placebo. If the patients who are given the drug improve more than the patients given the placebo by chance, it may appear that the drug is effective, but in fact the opposite is true.

By contrast, type II errors are errors of *omission*. In the example above, if the patients who got the drug did not get better at a higher rate than the ones who got the placebo and this was a random fluke, that would be a type II error.

Definition

Statistical Background

In statistical test theory, the notion of a statistical error is an integral part of hypothesis testing. This is conceptually similar to the judgement in a court trial. The null hypothesis corresponds to the position of the defendant: just as he is presumed to be innocent until proven guilty, so is the null hypothesis presumed to be true until the data provide convincing evidence against it. The alternative hypothesis corresponds to the position against the defendant. Specifically, the null hypothesis also involves the absence of a difference or the absence of an association. Thus, the null hypothesis can never be that there is a difference or an association.

If the result of the test corresponds with reality, then a correct decision has been made. However, if the result of the test does not correspond with reality, then an error has occurred. There are two situations in which the decision is wrong. The null hypothesis may be true, whereas we reject H_0. On the other hand, the alternative hypothesis H_1 may be true, whereas we do not reject H_0. Two types of error are distinguished: type I error and type II error.

Type I Error

The first kind of error is the mistaken rejection of a null hypothesis as the result of a test procedure. This kind of error is called a type I error (false positive) and is sometimes called an error of the first kind. In terms of the courtroom example, a type I error corresponds to convicting an innocent defendant.

Type II Error

The second kind of error is the mistaken failure to reject the null hypothesis as the result of a test procedure. This sort of error is called a type II error (false negative) and is also referred to as an error of the second kind. In terms of the courtroom example, a type II error corresponds to acquitting a criminal.

Crossover Error Rate

The crossover error rate (CER) is the point at which type I errors and type II errors are equal. A system with a lower CER value provides more accuracy than a system with a higher CER value.

False Positive and False Negative

In terms of false positives and false negatives, a positive result corresponds to rejecting the null hypothesis, while a negative result corresponds to failing to reject the null hypothesis; "false" means the conclusion drawn is incorrect. Thus, a type I error is equivalent to a false positive, and a type II error is equivalent to a false negative.

Error Rate

A perfect test would have zero false positives and zero false negatives. However, statistical methods are probabilistic, and it cannot be known for certain whether statistical conclusions are correct.

Whenever there is uncertainty, there is the possibility of making an error. Considering this, all statistical hypothesis tests have a probability of making type I and type II errors.

- The type I error rate is the probability of rejecting the null hypothesis given that it is true. The test is designed to keep the type I error rate below a prespecified bound called the significance level, usually denoted by the Greek letter α (alpha) and is also called the alpha level. Usually, the significance level is set to 0.05 (5%), implying that it is acceptable to have a 5% probability of incorrectly rejecting the true null hypothesis.
- The rate of the type II error is denoted by the Greek letter β (beta) and related to the power of a test, which equals $1-\beta$.

These two types of error rates are traded off against each other: for any given sample set, the effort to reduce one type of error generally results in increasing the other type of error.

The Quality of Hypothesis Test

The same idea can be expressed in terms of the rate of correct results and therefore used to minimize error rates and improve the quality of hypothesis test. To reduce the probability of committing a type I error, making the alpha value more stringent is both simple and efficient. To decrease the probability of committing a type II error, which is closely associated with analyses' power, either increasing the test's sample size or relaxing the alpha level could increase the analyses' power. A test statistic is robust if the type I error rate is controlled.

Varying different threshold (cut-off) values could also be used to make the test either more specific or more sensitive, which in turn elevates the test quality. For example, imagine a medical test, in which an experimenter might measure the concentration of a certain protein in the blood sample. The experimenter could adjust the threshold (black vertical line in the figure) and people would be diagnosed as having diseases if any number is detected above this certain threshold. According to the image, changing the threshold would result in changes in false positives and false negatives, corresponding to movement on the curve.

Application Domains

Medical Screening

Screening involves relatively cheap tests that are given to large populations, none of whom manifest any clinical indication of disease (e.g., Pap smears). Testing involves far more expensive, often invasive, procedures that are given only to those who manifest some clinical indication of disease, and are most often applied to confirm a suspected diagnosis.

For example, most states in the US require newborns to be screened for phenylketonuria and hypothyroidism, among other congenital disorders.

- **Hypothesis:** "The newborns have phenylketonuria and hypothyroidism".
- **Null Hypothesis (H_0):** "The newborns do not have phenylketonuria and hypothyroidism".
- **Type I Error (False Positive):** The true fact is that the newborns do not have phenylketonuria and hypothyroidism but we consider they have the disorders according to the data.
- **Type II Error (False Negative):** The true fact is that the newborns have phenylketonuria and hypothyroidism but we consider they do not have the disorders according to the data.

Although they display a high rate of false positives, the screening tests are considered valuable because they greatly increase the likelihood of detecting these disorders at a far earlier stage.

The simple blood tests used to screen possible blood donors for HIV and hepatitis have a significant rate of false positives; however, physicians use much more expensive and far more precise tests to determine whether a person is actually infected with either of these viruses.

Perhaps the most widely discussed false positives in medical screening come from the breast cancer screening procedure mammography. The US rate of false positive mammograms is up to 15%, the highest in world. One consequence of the high false positive rate in the US is that, in any 10-year period, half of the American women screened receive a false positive mammogram. False positive mammograms are costly, with over $100 million spent annually in the U.S. on follow-up testing and treatment. They also cause women unneeded anxiety. As a result of the high false positive rate in the US, as many as 90–95% of women who get a positive mammogram do not have the condition. The lowest rate in the world is in the Netherlands, 1%. The lowest rates are generally in Northern Europe where mammography films are read twice and a high threshold for additional testing is set (the high threshold decreases the power of the test).

The ideal population screening test would be cheap, easy to administer, and produce zero false negatives, if possible. Such tests usually produce more false positives, which can subsequently be sorted out by more sophisticated (and expensive) testing.

Medical Testing

False negatives and false positives are significant issues in medical testing.

- **Hypothesis:** "The patients have the specific disease".
- **Null Hypothesis (H_0):** "The patients do not have the specific disease".
- **Type I Error (False Positive):** The true fact is that the patients do not have a specific disease but the physician judges the patient is ill according to the test reports.
- **Type II Error (False Negative):** The true fact is that the disease is actually present but the test reports provide a falsely reassuring message to patients and physicians that the disease is absent.

False positives can also produce serious and counter-intuitive problems when the condition being searched for is rare, as in screening. If a test has a false positive rate of one in ten thousand, but only one in a million samples (or people) is a true positive, most of the positives detected by that test will be false. The probability that an observed positive result is a false positive may be calculated using Bayes' theorem.

False negatives produce serious and counter-intuitive problems, especially when the condition being searched for is common. If a test with a false negative rate of only 10% is used to test a population with a true occurrence rate of 70%, many of the negatives detected by the test will be false. This sometimes leads to inappropriate or inadequate treatment of both the patient and their disease. A common example is relying on cardiac stress tests to detect coronary atherosclerosis, even though cardiac stress tests are known to only detect limitations of coronary artery blood flow due to advanced stenosis.

Biometrics

Biometric matching, such as for fingerprint recognition, facial recognition or iris recognition, is susceptible to type I and type II errors.

- **Hypothesis:** "The input does not identify someone in the searched list of people".
- **Null Hypothesis:** "The input does identify someone in the searched list of people".
- **Type I Error (False Reject Rate):** The true fact is that the person is someone in the searched list but the system concludes that the person is not according to the data.
- **Type II Error (False Match Rate):** The true fact is that the person is not someone in the searched list but the system concludes that the person is someone whom we are looking for according to the data.

The probability of type I errors is called the "false reject rate" (FRR) or false non-match rate (FNMR), while the probability of type II errors is called the "false accept rate" (FAR) or false match rate (FMR).

If the system is designed to rarely match suspects then the probability of type II errors can be called the "false alarm rate". On the other hand, if the system is used for validation (and acceptance is the norm) then the FAR is a measure of system security, while the FRR measures user inconvenience level.

Security Screening

False positives are routinely found every day in airport security screening, which are ultimately visual inspection systems. The installed security alarms are intended to prevent weapons being brought onto aircraft; yet they are often set to such high sensitivity that they alarm many times a day for minor items, such as keys, belt buckles, loose change, mobile phones, and tacks in shoes.

- **Hypothesis:** "The item is a weapon".
- **Null Hypothesis:** "The item is not a weapon".
- **Type I Error (False Positive):** The true fact is that the item is not a weapon but the system still sounds an alarm.
- **Type II Error (False Negative):** The true fact is that the item is a weapon but the system keeps silent at this time.

The ratio of false positives (identifying an innocent traveler as a terrorist) to true positives (detecting a would-be terrorist) is, therefore, very high; and because almost every alarm is a false positive, the positive predictive value of these screening tests is very low. The relative cost of false results determines the likelihood that test creators allow these events to occur. As the cost of a false negative in this scenario is extremely high (not detecting a bomb being brought onto a plane could result in hundreds of deaths) whilst the cost of a false positive is relatively low (a reasonably simple further inspection) the most appropriate test is one with a low statistical specificity but high statistical sensitivity (one that allows a high rate of false positives in return for minimal false negatives).

Computers

The notions of false positives and false negatives have a wide currency in the realm of computers and computer applications, including computer security, spam filtering, malware, optical character recognition, and many others.

For example, in the case of spam filtering:

- **Hypothesis:** "The message is spam".
- **Null Hypothesis:** "The message is not spam".

- **Type I Error (False Positive):** Spam filtering or spam blocking techniques wrongly classify a legitimate email message as spam and, as a result, interfere with its delivery.
- **Type II Error (False Negative):** Spam email is not detected as spam, but is classified as non-spam.

While most anti-spam tactics can block or filter a high percentage of unwanted emails, doing so without creating significant false-positive results is a much more demanding task. A low number of false negatives is an indicator of the efficiency of spam filtering.

12.2.5 Common Statistical Tests

Different tests of significance are used depending on the type of data and the research question. Some common tests include:

T-Test

The t-test is used to compare the means of two groups and determine whether the difference between them is statistically significant. It is commonly applied in experimental studies where the goal is to compare the effects of different treatments or interventions.

- **Independent T-Test:** Compares the means of two independent groups (e.g., treatment vs. control group).
- **Paired T-Test:** Compares the means of two related groups (e.g., before and after treatment in the same group of patients).

ANOVA (Analysis of Variance)

ANOVA is used to compare the means of three or more groups to determine whether at least one group differs significantly from the others. ANOVA is widely used in experiments that involve multiple treatment groups or conditions.

Chi-Square Test

The chi-square test is used to examine the association between two categorical variables. It determines whether the observed frequencies in different categories differ from what would be expected by chance.

Z-Test

The z-test is used to compare sample and population means when the population variance is known. It is commonly used in large-sample studies to test hypotheses about population proportions or means.

12.3 CONFIDENCE INTERVALS AND HYPOTHESIS TESTING

Confidence intervals and hypothesis testing are closely related, as both are tools used to make inferences about a population. A confidence interval can also be used to test a hypothesis. If the value specified by the null hypothesis falls outside the confidence interval, the null hypothesis is rejected. For example, if a 95% confidence interval for the difference in means does not include zero, we can reject the null hypothesis that the two means are equal.

12.4 PRACTICAL APPLICATIONS OF TESTS OF SIGNIFICANCE AND ESTIMATION

In various fields, tests of significance and estimation are essential for making data-driven decisions. For instance, in medicine, significance testing is used to determine whether a new treatment is more effective than an existing one. In marketing, these methods help analyze consumer behavior, while in economics, they assist in assessing the impact of policies or interventions on economic indicators.

12.5 CONCLUSION

Tests of significance and estimation are powerful tools in statistical analysis, enabling researchers to draw meaningful conclusions from sample data. Estimation provides point estimates and confidence intervals, offering insights into population parameters, while tests of significance help assess the validity of hypotheses. By understanding and applying these methods, researchers and practitioners can make informed, data-driven decisions that stand up to scrutiny. These concepts form the foundation of much of modern research and are critical to fields that rely on statistical evidence.

CHAPTER 13
LINEAR REGRESSION AND CORRELATION

13.1 LINEAR REGRESSION

In statistics, linear regression is a statistical model that estimates the linear relationship between a scalar response (dependent variable) and one or more explanatory variables (regressor or independent variable). The case of one explanatory variable is called *simple linear regression*; for more than one, the process is called multiple linear regression. This term is distinct from multivariate linear regression, where multiple correlated dependent variables are predicted, rather than a single scalar variable. If the explanatory variables are measured with error then errors-in-variables models are required, also known as *measurement error models*.

In linear regression, the relationships are modeled using linear predictor functions whose unknown model parameters are estimated from the data. Such models are called linear models. Most commonly, the conditional mean of the response given the values of the explanatory variables (or predictors) is assumed to be an affine function of those values; less commonly, the conditional median or some other quantile is used. Like all forms of regression analysis, linear regression focuses on the conditional probability distribution of the response given the values of the predictors, rather than on the joint probability distribution of all of these variables, which is the domain of multivariate analysis.

Linear regression is also a type of machine-learning algorithm, more specifically a supervised machine-learning algorithm, that learns from the labelled datasets and maps the data points to the most optimized linear functions that can be used for prediction on new datasets.

Linear regression was the first type of regression analysis to be studied rigorously, and to be used extensively in practical applications. This is because models which depend linearly on their unknown parameters are easier to fit than models which are non-linearly related to their parameters and because the statistical properties of the resulting estimators are easier to determine.

Linear regression has many practical uses. Most applications fall into one of the following two broad categories:

- If the goal is error, i.e., variance reduction in prediction or forecasting, linear regression can be used to fit a predictive model to an observed data set of values of the response and explanatory variables. After developing such a model, if additional values of the explanatory variables are collected without an accompanying response value, the fitted model can be used to make a prediction of the response.
- If the goal is to explain variation in the response variable that can be attributed to variation in the explanatory variables, linear regression analysis can be applied to quantify the strength of the relationship between the response and the explanatory variables, and in particular to determine whether some explanatory variables may have no linear relationship with the response at all, or to identify which subsets of explanatory variables may contain redundant information about the response.

Linear regression models are often fitted using the least squares approach, but they may also be fitted in other ways, such as by minimizing the "lack of fit" in some other norm (as with least absolute deviations regression), or by minimizing a penalized version of the least squares cost function as in ridge regression (L^2-norm penalty) and lasso (L^1-norm penalty). Use of the Mean Squared Error (MSE) as the cost on a dataset that has many large outliers, can result in a model that fits the outliers more than the true data due to the higher importance assigned by MSE to large errors. So, cost functions that are robust to outliers should be used if the dataset has many large outliers. Conversely, the least squares approach can be used to fit models that are not linear models. Thus, although the terms "least squares" and "linear model" are closely linked, they are not synonymous.

13.1.1 Formulation

Given a data set $\{y_i, x_{i1}, \ldots, x_{ip}\}_{i=1}^{n}$ of n statistical units, a linear regression model assumes that the relationship between the dependent variable y and the vector of regressors x is linear. This relationship is modeled through a *disturbance term* or *error variable* ε—an unobserved random variable that adds "noise" to the linear relationship between the dependent variable and regressors. Thus the model takes the form,

$$y_i = \beta_0 + \beta_1 x_{i1} + \cdots + \beta_p x_{ip} + \varepsilon_i = \mathbf{x}_i^{\mathsf{T}} \boldsymbol{\beta} + \varepsilon_i, \qquad i = 1, \ldots, n,$$

Where ^{T} denotes the transpose, so that $x_i^{\mathsf{T}}\beta$ is the inner product between vectors x_i and β.

Often these n equations are stacked together and written in matrix notation as

$$\mathbf{y} = \mathbf{X}\boldsymbol{\beta} + \boldsymbol{\varepsilon},$$

Where

$$\mathbf{y} = \begin{bmatrix} y_1 \\ y_2 \\ \vdots \\ y_n \end{bmatrix},$$

$$\mathbf{X} = \begin{bmatrix} \mathbf{x}_1^{\mathsf{T}} \\ \mathbf{x}_2^{\mathsf{T}} \\ \vdots \\ \mathbf{x}_n^{\mathsf{T}} \end{bmatrix} = \begin{bmatrix} 1 & x_{11} & \cdots & x_{1p} \\ 1 & x_{21} & \cdots & x_{2p} \\ \vdots & \vdots & \ddots & \vdots \\ 1 & x_{n1} & \cdots & x_{np} \end{bmatrix},$$

$$\hat{\mathbf{a}} = \begin{bmatrix} \beta_0 \\ \beta_1 \\ \beta_2 \\ \vdots \\ \beta_p \end{bmatrix}, \quad \mathring{\mathbf{a}} = \begin{bmatrix} \varepsilon_1 \\ \varepsilon_2 \\ \vdots \\ \varepsilon_n \end{bmatrix}.$$

Assumptions

Standard linear regression models with standard estimation techniques make a number of assumptions about the predictor variables, the response variable and their relationship. Numerous extensions have been developed that allow each of these assumptions to be relaxed (i.e., reduced to a weaker form), and in some cases eliminated entirely. Generally these extensions make the estimation procedure more complex and time-consuming, and may also require more data in order to produce an equally precise model.

The following are the major assumptions made by standard linear regression models with standard estimation techniques (e.g., ordinary least squares):

- **Weak Exogeneity:** This essentially means that the predictor variables x can be treated as fixed values, rather than random variables. This means, for example, that the predictor variables are assumed to be error-free—that is, not contaminated with measurement errors. Although this assumption is not realistic in many settings, dropping it leads to significantly more difficult errors-in-variables models.
- **Linearity:** This means that the mean of the response variable is a linear combination of the parameters (regression coefficients) and the predictor variables. Note that this assumption is much less restrictive than it may at first seem. Because the predictor variables are treated as fixed values, linearity is really only a restriction on the parameters. The predictor variables themselves can be arbitrarily transformed, and in fact multiple copies of the same underlying predictor variable can be added, each one transformed differently. This technique is used, for

example, in polynomial regression, which uses linear regression to fit the response variable as an arbitrary polynomial function (up to a given degree) of a predictor variable. With this much flexibility, models such as polynomial regression often have "too much power", in that they tend to overfit the data. As a result, some kind of regularization must typically be used to prevent unreasonable solutions coming out of the estimation process. Common examples are ridge regression and lasso regression. Bayesian linear regression can also be used, which by its nature is more or less immune to the problem of overfitting. (In fact, ridge regression and lasso regression can both be viewed as special cases of Bayesian linear regression, with particular types of prior distributions placed on the regression coefficients.)

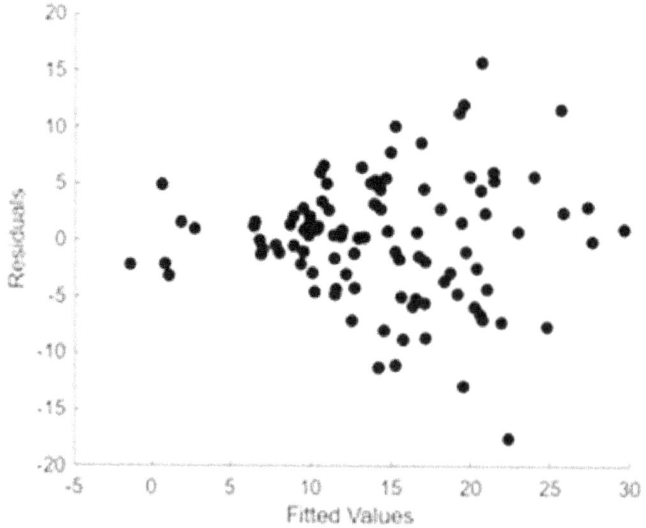

Fig 1: Visualization of Heteroscedasticity in a Scatter Plot against 100 Random Fitted Values Using Matlab

- **Constant Variance:** This means that the variance of the errors does not depend on the values of the predictor variables. Thus the variability of the responses for given fixed values of the predictors is the same regardless of how large or small the responses are. This is often not the case, as a variable whose mean is large will typically have a greater variance than one whose mean is small. For example, a person whose income is predicted to be \$100,000 may easily have an actual income of \$80,000 or \$120,000—i.e., a standard deviation of around \$20,000—while another person with a predicted income of \$10,000 is unlikely to have the same \$20,000 standard deviation, since that would imply their actual income could vary anywhere between -\$10,000 and \$30,000. (In fact, as this shows, in many cases—often the same cases where the assumption of normally distributed errors fails—the variance or standard deviation should be predicted to be proportional to the mean, rather than constant.) The absence of homoscedasticity is called heteroscedasticity. In order to check this assumption, a plot of residuals versus predicted values (or the values of each individual predictor) can be examined for a "fanning effect" (i.e., increasing or decreasing vertical spread as one moves left to right on the plot). A plot of the absolute or squared residuals versus the predicted values (or each predictor) can also be examined for a trend or curvature. Formal tests can also be used. The presence of heteroscedasticity will result in an overall "average" estimate of variance being used instead of one that takes into account the true variance structure. This leads to less precise (but in the case of ordinary least squares, not biased) parameter estimates and biased standard errors, resulting in misleading tests and interval estimates. The mean squared error for the model will also be wrong. Various estimation techniques including weighted least squares and the use of heteroscedasticity-consistent standard errors can handle heteroscedasticity in a quite general way. Bayesian linear regression techniques can also be used when the variance is assumed to be a function of the mean. It is also possible in some cases to fix the problem by applying a transformation to the response variable (e.g., fitting the logarithm of the response variable using a linear regression model, which implies that the response variable itself has a log-normal distribution rather than a normal distribution).

- **Independence of Errors:** This assumes that the errors of the response variables are uncorrelated with each other. (Actual statistical independence is a stronger condition than mere lack of correlation and is often not needed, although it can be exploited if it is known to hold.) Some methods such as generalized least squares are capable of handling correlated errors, although they typically require significantly more data unless some sort of regularization is used to bias the model towards assuming uncorrelated errors. Bayesian linear regression is a general way of handling this issue.
- **Lack of Perfect Multicollinearity in the Predictors:** For standard least squares estimation methods, the design matrix X must have full column rank p; otherwise perfect multicollinearity exists in the predictor variables, meaning a linear relationship exists between two or more predictor variables. This can be caused by accidentally duplicating a

variable in the data, using a linear transformation of a variable along with the original (e.g., the same temperature measurements expressed in Fahrenheit and Celsius), or including a linear combination of multiple variables in the model, such as their mean. It can also happen if there is too little data available compared to the number of parameters to be estimated (e.g., fewer data points than regression coefficients). Near violations of this assumption, where predictors are highly but not perfectly correlated, can reduce the precision of parameter estimates. In the case of perfect multicollinearity, the parameter vector β will be non-identifiable—it has no unique solution. In such a case, only some of the parameters can be identified (i.e., their values can only be estimated within some linear subspace of the full parameter space R^p). Methods for fitting linear models with multicollinearity have been developed, some of which require additional assumptions such as "effect sparsity"—that a large fraction of the effects are exactly zero. Note that the more computationally expensive iterated algorithms for parameter estimation, such as those used in generalized linear models, do not suffer from this problem.

- **Assumption of Zero Mean of Residuals:** In regression analysis, another critical assumption is that the mean of the residuals is zero or close to zero. This assumption is fundamental for the validity of any conclusions drawn from the least squares estimates of the parameters. Residuals are the differences between the observed values and the values predicted by the model. If the mean of these residuals is not zero, it implies that the model consistently overestimates or underestimates the observed values, indicating a potential bias in the model estimation. Ensuring that the mean of the residuals is zero allows the model to be considered unbiased in terms of its error, which is crucial for the accurate interpretation of the regression coefficients.

Violations of these assumptions can result in biased estimations of β, biased standard errors, untrustworthy confidence intervals and significance tests.

Beyond these assumptions, several other statistical properties of the data strongly influence the performance of different estimation methods:

- The statistical relationship between the error terms and the regressors plays an important role in determining whether an estimation procedure has desirable sampling properties such as being unbiased and consistent.
- The arrangement, or probability distribution of the predictor variables x has a major influence on the precision of estimates of β. Sampling and

design of experiments are highly developed subfields of statistics that provide guidance for collecting data in such a way to achieve a precise estimate of β.

Interpretation

A fitted linear regression model can be used to identify the relationship between a single predictor variable x_j and the response variable y when all the other predictor variables in the model are "held fixed". Specifically, the interpretation of β_j is the expected change in y for a one-unit change in x_j when the other covariates are held fixed—that is, the expected value of the partial derivative of y with respect to x_j. This is sometimes called the *unique effect* of x_j on y. In contrast, the *marginal effect* of x_j on y can be assessed using a correlation coefficient or simple linear regression model relating only x_j to y; this effect is the total derivative of y with respect to x_j.

Care must be taken when interpreting regression results, as some of the regressors may not allow for marginal changes (such as dummy variables, or the intercept term), while others cannot be held fixed (recall the example from the introduction: it would be impossible to "hold t_i fixed" and at the same time change the value of t_i^2).

It is possible that the unique effect be nearly zero even when the marginal effect is large. This may imply that some other covariate captures all the information in x_j, so that once that variable is in the model, there is no contribution of x_j to the variation in y. Conversely, the unique effect of x_j can be large while its marginal effect is nearly zero. This would happen if the other covariates explained a great deal of the variation of y, but they mainly explain variation in a way that is complementary to what is captured by x_j. In this case, including the other variables in the model reduces the part of the variability of y that is unrelated to x_j, thereby strengthening the apparent relationship with x_j.

The meaning of the expression "held fixed" may depend on how the values of the predictor variables arise. If the experimenter directly sets the values of the predictor variables according to a study design, the comparisons of interest may literally correspond to comparisons among units whose predictor variables have been "held fixed" by the experimenter. Alternatively, the expression "held fixed" can refer to a selection that takes place in the context of data analysis. In this case, we "hold a variable fixed" by restricting our attention to the subsets of the data that happen to have a common value for the given predictor variable. This is the only interpretation of "held fixed" that can be used in an observational study.

The notion of a "unique effect" is appealing when studying a complex system where multiple interrelated components influence the response variable. In some cases, it can literally be interpreted as the causal effect of an intervention that is linked to the value of a predictor variable. However, it has been argued that in many cases multiple regression analysis fails to clarify the relationships between the predictor variables and the response variable when the predictors are correlated with each other and are not assigned following a study design.

13.1.2 Extensions

Numerous extensions of linear regression have been developed, which allow some or all of the assumptions underlying the basic model to be relaxed.

Simple and Multiple Linear Regression

The simplest case of a single scalar predictor variable x and a single scalar response variable y is known as *simple linear regression.* The extension to multiple and/or vector-valued predictor variables (denoted with a capital X) is known as multiple linear regression, also known as multivariable linear regression (not to be confused with *multivariate linear regression*).

Multiple linear regression is a generalization of simple linear regression to the case of more than one independent variable, and a special case of general linear models, restricted to one dependent variable. The basic model for multiple linear regression is $Y_i = \beta_0 + \beta_1 X_{i1} + \beta_2 X_{i2} + \ldots + \beta_p X_{ip} + \epsilon_i$ for each observation $i = 1, \ldots, n$.

In the formula above we consider *n* observations of one dependent variable and *p* independent variables. Thus, Y_i is the ith observation of the dependent variable, X_{ij} is i^{th} observation of the jth independent variable, $j = 1, 2, ..., p$. The values β_j represent parameters to be estimated, and ε_i is the ith independent identically distributed normal error.

In the more general multivariate linear regression, there is one equation of the above form for each of $m > 1$ dependent variables that share the same set of explanatory variables and hence are estimated simultaneously with each other:

$$Y_{ij} = \beta_{0j} + \beta_{1j} X_{i1} + \beta_{2j} X_{i2} + \ldots + \beta_{pj} X_{ip} + \epsilon_{ij}$$

For all observations indexed as $i = 1, \ldots, n$ and for all dependent variables indexed as $j = 1, \ldots,$ m.

Nearly all real-world regression models involve multiple predictors, and basic descriptions of linear regression are often phrased in terms of the multiple regression model. Note, however, that in these cases the response variable *y* is still a scalar. Another term, *multivariate linear regression*, refers to cases where *y* is a vector, i.e., the same as *general linear regression.*

General Linear Models

The general linear model considers the situation when the response variable is not a scalar (for each observation) but a vector, y_i. Conditional linearity of

$E(\mathbf{y} \mid \mathbf{x}_i) = \mathbf{x}_i^{\mathsf{T}} B$ is still assumed, with a matrix B replacing the vector β of the classical linear regression model. Multivariate analogues of ordinary least squares (OLS) and generalized least squares (GLS) have been developed. "General linear models" are also called "multivariate linear models". These are not the same as multivariable linear models (also called "multiple linear models").

Heteroscedastic Models

Various models have been created that allow for heteroscedasticity, i.e. the errors for different response variables may have different variances. For example, weighted least squares is a method for estimating linear regression models when the response variables may have different error variances, possibly with correlated errors. Heteroscedasticity-consistent standard errors is an improved method for use with uncorrelated but potentially heteroscedastic errors.

Generalized Linear Models

The Generalized linear model (GLM) is a framework for modeling response variables that are bounded or discrete. This is used, for example:

- when modeling positive quantities (e.g., prices or populations) that vary over a large scale—which are better described using a skewed distribution such as the log-normal distribution or Poisson distribution (although GLMs are not used for log-normal data, instead the response variable is simply transformed using the logarithm function);
- when modeling categorical data, such as the choice of a given candidate in an election (which is better described using a Bernoulli distribution/ binomial distribution for binary choices, or a categorical distribution/ multinomial distribution for multi-way choices), where there are a fixed number of choices that cannot be meaningfully ordered; and
- when modeling ordinal data, e.g. ratings on a scale from 0 to 5, where the different outcomes can be ordered but where the quantity itself may not have any absolute meaning (e.g. a rating of 4 may not be "twice as good" in any objective sense as a rating of 2, but simply indicates that it is better than 2 or 3 but not as good as 5).

Generalized linear models allow for an arbitrary *link function*, g, that relates the mean of the response variable(s) to the predictors: $E(Y) = g^{-1}(XB)$. The link function is often related to the distribution of the response, and in particular it typically has the effect of transforming between the $(-\infty, \infty)$ range of the linear predictor and the range of the response variable.

Some common examples of GLMs are:

- Poisson regression for count data.

- Logistic regression and probit regression for binary data.
- Multinomial logistic regression and multinomial probit regression for categorical data.
- Ordered logit and ordered probit regression for ordinal data.

Single index models allow some degree of nonlinearity in the relationship between x and y, while preserving the central role of the linear predictor $\beta'x$ as in the classical linear regression model. Under certain conditions, simply applying OLS to data from a single-index model will consistently estimate β up to a proportionality constant.

Hierarchical Linear Models

Hierarchical linear models (or *multilevel regression*) organizes the data into a hierarchy of regressions, for example where A is regressed on B, and B is regressed on C. It is often used where the variables of interest have a natural hierarchical structure such as in educational statistics, where students are nested in classrooms, classrooms are nested in schools, and schools are nested in some administrative grouping, such as a school district.

The response variable might be a measure of student achievement such as a test score, and different covariates would be collected at the classroom, school, and school district levels.

Errors-in-Variables

Errors-in-variables models (or "measurement error models") extend the traditional linear regression model to allow the predictor variables X to be observed with error. This error causes standard estimators of β to become biased. Generally, the form of bias is an attenuation, meaning that the effects are biased toward zero.

13.1.3 Applications

Linear regression is widely used in biological, behavioral and social sciences to describe possible relationships between variables. It ranks as one of the most important tools used in these disciplines.

Trend Line

A trend line represents a trend, the long-term movement in time series data after other components have been accounted for. It tells whether a particular data set (say GDP, oil prices or stock prices) have increased or decreased over the period of time. A trend line could simply be drawn by eye through a set of data points, but more properly their position and slope is calculated using statistical techniques like linear regression. Trend lines typically are straight lines, although some variations use higher degree polynomials depending on the degree of curvature desired in the line. Trend lines are sometimes used in business analytics to show changes in data over time. This has the advantage of being simple. Trend lines are often used to argue that a particular action or event (such as training, or an advertising campaign) caused observed changes at a point in time. This is a simple technique, and does not require a control group, experimental design, or a sophisticated analysis technique. However, it suffers from a lack of scientific validity in cases where other potential changes can affect the data.

Epidemiology

Early evidence relating tobacco smoking to mortality and morbidity came from observational studies employing regression analysis. In order to reduce spurious correlations when analyzing observational data, researchers usually include several variables in their regression models in addition to the variable of primary interest.

For example, in a regression model in which cigarette smoking is the independent variable of primary interest and the dependent variable is lifespan measured in years, researchers might include education and income as additional independent variables, to ensure that any observed effect of smoking on lifespan is not due to those other socio-economic factors. However, it is never possible to include all possible confounding variables in an empirical analysis. For example, a hypothetical gene might increase mortality and also cause people to smoke more. For this reason, randomized controlled trials are often able to generate more compelling evidence of causal relationships than can be obtained using regression analyses of observational data. When controlled experiments are not feasible, variants of regression analysis such as instrumental variables regression may be used to attempt to estimate causal relationships from observational data.

Finance

The capital asset pricing model uses linear regression as well as the concept of beta for analyzing and quantifying the systematic risk of an investment. This comes directly from the beta coefficient of the linear regression model that relates the return on the investment to the return on all risky assets.

Economics

Linear regression is the predominant empirical tool in economics. For example, it is used to predict consumption spending, fixed investment spending, inventory investment, purchases of a country's exports, spending on imports, the demand to hold liquid assets, labor demand, and labor supply.

Environmental Science

Linear regression finds application in a wide range of environmental science applications such as land use, infectious diseases, and air pollution. For example, linear regression can be used to predict the changing effects of car pollution.

Building Science

Linear regression is commonly used in building science field studies to derive characteristics of building occupants. In a thermal comfort field study, building scientists usually ask occupants' thermal sensation votes, which range from -3 (feeling cold) to 0 (neutral) to +3 (feeling hot), and measure occupants' surrounding temperature data. A neutral or comfort temperature can be calculated based on a linear regression between the thermal sensation vote and indoor temperature, and setting the thermal sensation vote as zero. However, there has been a debate on the regression direction: regressing thermal sensation votes (y-axis) against indoor temperature (x-axis) or the opposite: regressing indoor temperature (y-axis) against thermal sensation votes (x-axis).

Machine Learning

Linear regression plays an important role in the subfield of artificial intelligence known as machine learning. The linear regression algorithm is one of the fundamental supervised machine-learning algorithms due to its relative simplicity and well-known properties.

13.2 CORRELATION

In statistics, correlation or dependence is any statistical relationship, whether causal or not, between two random variables or bivariate data. Although in the broadest sense, "correlation" may indicate any type of association, in statistics it usually refers to the degree to which a pair of variables are *linearly* related. Familiar examples of dependent phenomena include the correlation between the height of parents and their offspring, and the correlation between the price of a good and the quantity the consumers are willing to purchase, as it is depicted in the so-called demand curve.

Correlations are useful because they can indicate a predictive relationship that can be exploited in practice. For example, an electrical utility may produce less power on a mild day based on the correlation between electricity demand and weather. In this example, there is a causal relationship, because extreme weather causes people to use more electricity for heating or cooling. However, in general, the presence of a correlation is not sufficient to infer the presence of a causal relationship (i.e., correlation does not imply causation).

Formally, random variables are *dependent* if they do not satisfy a mathematical property of probabilistic independence. In informal parlance, *correlation* is synonymous with *dependence*. However, when used in a technical sense, correlation refers to any of several specific types of mathematical relationship between the conditional expectation of one variable given the other is not constant as the conditioning variable changes; broadly correlation in this specific sense is used when $E(Y \mid X = x)$ is related to x x in some manner (such as linearly, monotonically, or perhaps according to some particular functional form such as logarithmic). Essentially, correlation is the measure of how two or more variables are related to one another.

There are several correlation coefficients, often denoted $\rho\, or\, r$, measuring the degree of correlation. The most common of these is the *Pearson correlation coefficient*, which is sensitive only to a linear relationship between two variables (which may be present even when one variable is a nonlinear function of the other). Other correlation coefficients – such as *Spearman's rank correlation* – have been developed to be more robust than Pearson›s, that is, more sensitive to nonlinear relationships. Mutual information can also be applied to measure dependence between two variables.

13.2.1 Pearson's Product-Moment Coefficient

The most familiar measure of dependence between two quantities is the Pearson product-moment correlation coefficient (PPMCC), or "Pearson's correlation coefficient", commonly called simply "the correlation coefficient". It is obtained by taking the ratio of the covariance of the two variables in question of our numerical dataset, normalized to the square root of their variances.

Mathematically, one simply divides the covariance of the two variables by the product of their standard deviations. Karl Pearson developed the coefficient from a similar but slightly different idea by Francis Galton.

A Pearson product-moment correlation coefficient attempts to establish a line of best fit through a dataset of two variables by essentially laying out the expected values and the resulting Pearson's correlation coefficient indicates how far away the actual dataset is from the expected values.

Depending on the sign of our Pearson's correlation coefficient, we can end up with either a negative or positive correlation if there is any sort of relationship between the variables of our data set.

The population correlation coefficient $\rho_{X,Y}$ between two random variables X and Y with expected values $\mu_X\, and\, \mu_Y$ and standard deviations $\sigma_X\, and\, \sigma_Y$ is defined as:

$$\rho_{X,Y} = \mathrm{corr}(X,Y) = \frac{\mathrm{cov}(X,Y)}{\sigma_X \sigma_Y} = \frac{\mathrm{E}[(X-\mu_X)(Y-\mu_Y)]}{\sigma_X \sigma_Y}, \quad \text{if } \sigma_X \sigma_Y > 0.$$

The Pearson correlation is defined only if both standard deviations are finite and positive. An alternative formula purely in terms of moments is:

$$\rho_{X,Y} = \frac{E(XY) - E(X)E(Y)}{\sqrt{E(X^2) - E(X)^2} \cdot \sqrt{E(Y^2) - E(Y)^2}}$$

Correlation and Independence

It is a corollary of the Cauchy–Schwarz inequality that the absolute value of the Pearson correlation coefficient is not bigger than 1. Therefore, the value of a correlation coefficient ranges between -1 and +1. The correlation coefficient is +1 in the case of a perfect direct (increasing) linear relationship (correlation), -1 in the case of a perfect inverse (decreasing) linear relationship (anti-correlation), and some value in the open interval $(-1,1)$ in all other cases, indicating the degree of linear dependence between the variables. As it approaches zero there is less of a relationship (closer to uncorrelated). The closer the coefficient is to either -1 or 1, the stronger the correlation between the variables.

If the variables are independent, Pearson's correlation coefficient is 0. However, because the correlation coefficient detects only linear dependencies between two variables, the converse is not necessarily true. A correlation coefficient of 0 does not imply that the variables are independent.

$$\begin{aligned} &X,Y \text{ independent} && \Rightarrow \rho_{X,Y} = 0 \quad (X,Y \text{ uncorrelated}) \\ &\rho_{X,Y} = 0 \ (X,Y \text{ uncorrelated}) && \nRightarrow X,Y \text{ independent} \end{aligned}$$

13.2.2 Other Measures of Dependence Among Random Variables

The information given by a correlation coefficient is not enough to define the dependence structure between random variables. The correlation coefficient completely defines the dependence structure only in very particular cases, for example when the distribution is a multivariate normal distribution. In the case of elliptical distributions it characterizes the (hyper-)ellipses of equal density; however, it does not completely characterize the dependence structure (for example, a multivariate t-distribution's degrees of freedom determine the level of tail dependence).

For continuous variables, multiple alternative measures of dependence were introduced to address the deficiency of Pearson's correlation that it can be zero for dependent random variables. They all share the important property that a value of zero implies independence.

This led some authors to recommend their routine usage, particularly of Distance correlation. Another alternative measure is the Randomized Dependence Coefficient. The RDC is a computationally efficient, copula-based measure of dependence between multivariate random variables and is invariant with respect to non-linear scalings of random variables.

One important disadvantage of the alternative, more general measures is that, when used to test whether two variables are associated, they tend to have lower power compared to Pearson's correlation when the data follow a multivariate normal distribution. This is an implication of the No free lunch theorem theorem. To detect all kinds of relationships, these measures have to sacrifice power on other relationships, particularly for the important special case of a linear relationship with Gaussian marginals, for which Pearson's correlation is optimal. Another problem concerns interpretation. While Person's correlation can be interpreted for all values, the alternative measures can generally only be interpreted meaningfull at the extremes.

For two binary variables, the odds ratio measures their dependence, and takes range non-negative numbers, possibly infinity: $[0,+\infty]$. Related statistics such as Yule's Y and Yule's Q normalize this to the correlation-like range $[-1,1]$. The odds ratio is generalized by the logistic model to model cases where the dependent variables are discrete and there may be one or more independent variables.

The correlation ratio, entropy-based mutual information, total correlation, dual total correlation and polychoric correlation are all also capable of detecting more general dependencies, as is consideration of the copula between them, while the coefficient of determination generalizes the correlation coefficient to multiple regression.

13.2.3 Sensitivity to the Data Distribution

The degree of dependence between variables X and Y does not depend on the scale on which the variables are expressed.

That is, if we are analyzing the relationship between X and Y, most correlation measures are unaffected by transforming X to $a + bX$ and Y to $c + dY$, where a, b, c, and d are constants (b and d being positive). This is true of some correlation statistics as well as their population analogues. Some correlation statistics, such as the rank correlation coefficient, are also invariant to monotone transformations of the marginal distributions of X and/or Y.

Most correlation measures are sensitive to the manner in which X and Y are sampled. Dependencies tend to be stronger if viewed over a wider range of values. Thus, if we consider the correlation coefficient between the heights of fathers and their sons over all adult males, and compare it to the same correlation coefficient calculated when the fathers are selected to be between 165 cm and 170 cm in height, the correlation will be weaker in the latter case.

Several techniques have been developed that attempt to correct for range restriction in one or both variables, and are commonly used in meta-analysis; the most common are Thorndike's case II and case III equations.

Various correlation measures in use may be undefined for certain joint distributions of X and Y. For example, the Pearson correlation coefficient is defined in terms of moments, and hence will be undefined if the moments are undefined. Measures of dependence based on quantiles are always defined. Sample-based statistics intended to estimate population measures of dependence may or may not have desirable statistical properties such as being unbiased, or asymptotically consistent, based on the spatial structure of the population from which the data were sampled.

Sensitivity to the data distribution can be used to an advantage. For example, scaled correlation is designed to use the sensitivity to the range in order to pick out correlations between fast components of time series. By reducing the range of values in a controlled manner, the correlations on long time scale are filtered out and only the correlations on short time scales are revealed.

13.2.4 Common Misconceptions

Correlation and Causality

The conventional dictum that "correlation does not imply causation" means that correlation cannot be used by itself to infer a causal relationship between the variables. This dictum should not be taken to mean that correlations cannot indicate the potential existence of causal relations. However, the causes underlying the correlation, if any, may be indirect and unknown, and high correlations also overlap with identity relations (tautologies), where no causal process exists. Consequently, a correlation between two variables is not a sufficient condition to establish a causal relationship (in either direction).

CHAPTER 14

THE CHI-SQUARE TEST

A chi-squared test (also chi-square or χ^2 test) is a statistical hypothesis test used in the analysis of contingency tables when the sample sizes are large. In simpler terms, this test is primarily used to examine whether two categorical variables (*two dimensions of the contingency table*) are independent in influencing the test statistic (*values within the table*). The test is valid when the test statistic is chi-squared distributed under the null hypothesis, specifically Pearson's chi-squared test and variants thereof. Pearson's chi-squared test is used to determine whether there is a statistically significant difference between the expected frequencies and the observed frequencies in one or more categories of a contingency table. For contingency tables with smaller sample sizes, a Fisher's exact test is used instead.

In the standard applications of this test, the observations are classified into mutually exclusive classes. If the null hypothesis that there are no differences between the classes in the population is true, the test statistic computed from the observations follows a χ^2 frequency distribution. The purpose of the test is to evaluate how likely the observed frequencies would be assuming the null hypothesis is true.

Test statistics that follow a χ^2 distribution occur when the observations are independent. There are also χ^2 tests for testing the null hypothesis of independence of a pair of random variables based on observations of the pairs.

Chi-squared tests often refers to tests for which the distribution of the test statistic approaches the χ^2 distribution asymptotically, meaning that the sampling distribution (if the null hypothesis is true) of the test statistic approximates a chi-squared distribution more and more closely as sample sizes increase.

14.1 HISTORY

In the 19th century, statistical analytical methods were mainly applied in biological data analysis and it was customary for researchers to assume that observations followed a normal distribution, such as Sir George Airy and Mansfield Merriman, whose works were criticized by Karl Pearson in his 1900 paper.

At the end of the 19th century, Pearson noticed the existence of significant skewness within some biological observations. In order to model the observations regardless of being normal or skewed, Pearson, in a series of articles published from 1893 to 1916, devised the Pearson distribution, a family of continuous probability distributions, which includes the normal distribution and many skewed distributions, and proposed a method of statistical analysis consisting of using the Pearson distribution to model the observation and performing a test of goodness of fit to determine how well the model really fits to the observations.

14.1.1 Pearson's Chi-Squared Test

In 1900, Pearson published a paper on the χ^2 test which is considered to be one of the foundations of modern statistics. In this paper, Pearson investigated a test of goodness of fit.

Suppose that n observations in a random sample from a population are classified into k mutually exclusive classes with respective observed numbers of observations x_i (for $i = 1,2,...,k$), and a null hypothesis gives the probability p_i that an observation falls into the ith class. So we have the expected numbers $m_i = np_i$ for all i, where

$$\sum_{i=1}^{k} p_i = 1$$

$$\sum_{i=1}^{k} m_i = n\sum_{i=1}^{k} p_i = n$$

Pearson proposed that, under the circumstance of the null hypothesis being correct, as $n \to \infty$ the limiting distribution of the quantity given below is the χ^2 distribution.

$$X^2 = \sum_{i=1}^{k} \frac{(x_i - m_i)^2}{m_i} = \sum_{i=1}^{k} \frac{x_i^2}{m_i} - n$$

Pearson dealt first with the case in which the expected numbers m_i are large enough known numbers in all cells assuming every observation x_i may be taken as normally distributed, and reached the result that, in the limit as n becomes large, X^2 follows the χ^2 distribution with $k - 1$ degrees of freedom.

However, Pearson next considered the case in which the expected numbers depended on the parameters that had to be estimated from the sample, and suggested that, with the notation of m_i being the true expected numbers and m'_i being the estimated expected numbers, the difference

$$X^2 - X'^2 = \sum_{i=1}^{k} \frac{x_i^2}{m_i} - \sum_{i=1}^{k} \frac{x_i^2}{m'_i}$$

will usually be positive and small enough to be omitted. In a conclusion, Pearson argued that if we regarded X'^2 as also distributed as χ^2 distribution with $k - 1$ degrees of freedom, the error in this approximation would not affect practical decisions. This conclusion caused some controversy in practical applications and was not settled for 20 years until Fisher's 1922 and 1924 papers.

14.2 OTHER EXAMPLES OF CHI-SQUARED TESTS

One test statistic that follows a chi-squared distribution exactly is the test that the variance of a normally distributed population has a given value based on a sample variance. Such tests are uncommon in practice because the true variance of the population is usually unknown. However, there are several statistical tests where the chi-squared distribution is approximately valid:

14.2.1 Fisher's Exact Test

For an exact test used in place of the 2 × 2 chi-squared test for independence.

14.2.2 Binomial Test

For an exact test used in place of the 2 × 1 chi-squared test for goodness of fit.

14.2.3 Other Chi-Squared Tests

- Cochran–Mantel–Haenszel chi-squared test.
- McNemar's test, used in certain 2×2 tables with pairing.
- Tukey's test of additivity.
- The portmanteau test in time-series analysis, testing for the presence of autocorrelation.
- Likelihood-ratio tests in general statistical modelling, for testing whether there is evidence of the need to move from a simple model to a more complicated one (where the simple model is nested within the complicated one).

14.3 YATES'S CORRECTION FOR CONTINUITY

Using the chi-squared distribution to interpret Pearson's chi-squared statistic requires one to assume that the discrete probability of observed binomial frequencies in the table can be approximated by the continuous chi-squared distribution. This assumption is not quite correct and introduces some error.

To reduce the error in approximation, Frank Yates suggested a correction for continuity that adjusts the formula for Pearson's chi-squared test by subtracting 0.5 from the absolute difference between each observed value and its expected value in a 2 × 2 contingency table. This reduces the chi-squared value obtained and thus increases its *p*-value.

	S	F	
A	a	b	a + b
B	c	d	c + d
	a + c	b + d	N

14.4 CHI-SQUARED TEST FOR VARIANCE IN A NORMAL POPULATION

If a sample of size n is taken from a population having a normal distribution, then there is a result which allows a test to be made of whether the variance of the population has a pre-determined value. For example, a manufacturing process might have been in stable condition for a long period, allowing a value for the variance to be determined essentially without error. Suppose that a variant of the process is being tested, giving rise to a small sample of n product items whose variation is to be tested. The test statistic T in this instance could be set to be the sum of squares about the sample mean, divided by the nominal value for the variance (i.e. the value to be tested as holding). Then T has a chi-squared distribution with n - 1 degrees of freedom. For example, if the sample size is 21, the acceptance region for T with a significance level of 5% is between 9.59 and 34.17.

14.5 EXAMPLE CHI-SQUARED TEST FOR CATEGORICAL DATA

Suppose there is a city of 1,000,000 residents with four neighborhoods: A, B, C, and D. A random sample of 650 residents of the city is taken and their occupation is recorded as "white collar", "blue collar", or "no collar". The null hypothesis is that each person's neighborhood of residence is independent of the person's occupational classification. The data are tabulated as:

	A	B	C	D	Total
White collar	90	60	104	95	349
Blue collar	30	50	51	20	151
No collar	30	40	45	35	150
Total	150	150	200	150	650

Let us take the sample living in neighborhood A, 150, to estimate what proportion of the whole 1,000,000 live in neighborhood A. Similarly we take 349/650 to estimate what proportion of the 1,000,000 are white-collar workers. By the assumption of independence under the hypothesis we should "expect" the number of white-collar workers in neighborhood A to be

$$150 \times \frac{349}{650} \approx 80.54$$

Then in that "cell" of the table, we have

$$\frac{(\text{observed} - \text{expected})^2}{\text{expected}} = \frac{(90 - 80.54)^2}{80.54} \approx 1.11$$

The sum of these quantities over all of the cells is the test statistic; in this case, ≈ 24.57. Under the null hypothesis, this sum has approximately a chi-squared distribution whose number of degrees of freedom is

$$(\text{number of rows} - 1)(\text{number of columns} - 1) = (3-1)(4-1) = 6$$

If the test statistic is improbably large according to that chi-squared distribution, then one rejects the null hypothesis of independence. Here we have a chi-squared value of 24.57, which is quite large, and therefore we have some evidence to reject the null hypothesis(H0). This means each person's neighbourhood of residence is correlated to the person's occupational classification.

A related issue is a test of homogeneity. Suppose that instead of giving every resident of each of the four neighborhoods an equal chance of inclusion in the sample, we decide in advance how many residents of each neighborhood to include. Then each resident has the same chance of being chosen as do all residents of the same neighborhood, but residents of different neighborhoods would have different probabilities of being chosen if the four sample sizes are not proportional to the populations of the four neighborhoods. In such a case, we would be testing "homogeneity" rather than "independence". The question is whether the proportions of blue-collar, white-collar, and no-collar workers in the four neighborhoods are the same. However, the test is done in the same way.

14.6 APPLICATIONS

In cryptanalysis, the chi-squared test is used to compare the distribution of plaintext and (possibly) decrypted ciphertext. The lowest value of the test means that the decryption was successful with high probability. This method can be generalized for solving modern cryptographic problems.

In bioinformatics, the chi-squared test is used to compare the distribution of certain properties of genes (e.g., genomic content, mutation rate, interaction network clustering, etc.) belonging to different categories (e.g., disease genes, essential genes, genes on a certain chromosome etc.).

CHAPTER 15
ANALYSIS OF VARIANCE

Analysis of variance (ANOVA) is a collection of statistical models and their associated estimation procedures (such as the "variation" among and between groups) used to analyze the differences among means. ANOVA was developed by the statistician Ronald Fisher. ANOVA is based on the law of total variance, where the observed variance in a particular variable is partitioned into components attributable to different sources of variation. In its simplest form, ANOVA provides a statistical test of whether two or more population means are equal, and therefore generalizes the *t*-test beyond two means. In other words, the ANOVA is used to test the difference between two or more means.

15.1 HISTORY

While the analysis of variance reached fruition in the 20th century, antecedents extend centuries into the past according to Stigler. These include hypothesis testing, the partitioning of sums of squares, experimental techniques and the additive model. Laplace was performing hypothesis testing in the 1770s. Around 1800, Laplace and Gauss developed the least-squares method for combining observations, which improved upon methods then used in astronomy and geodesy. It also initiated much study of the contributions to sums of squares. Laplace knew how to estimate a variance from a residual (rather than a total) sum of squares. By 1827, Laplace was using least squares methods to address ANOVA problems regarding measurements of atmospheric tides. Before 1800, astronomers had isolated observational errors resulting from reaction times (the "personal equation") and had developed methods of reducing the errors. The experimental methods used in the study of the personal equation were later accepted by the emerging field of psychology which developed strong (full factorial) experimental methods to which randomization and blinding were soon added. An eloquent non-mathematical explanation of the additive effects model was available in 1885.

Ronald Fisher introduced the term variance and proposed its formal analysis in a 1918 article on theoretical population genetics, *The Correlation Between Relatives on the Supposition of Mendelian Inheritance.* His first application of the analysis of variance to data analysis was published in 1921, *Studies in Crop Variation I.* This divided the variation of a time series into components representing annual causes and slow deterioration. Fisher's next piece, *Studies in Crop Variation II*, written with

Winifred Mackenzie and published in 1923, studied the variation in yield across plots sown with different varieties and subjected to different fertiliser treatments. Analysis of variance became widely known after being included in Fisher's 1925 book *Statistical Methods for Research Workers*. Randomization models were developed by several researchers. The first was published in Polish by Jerzy Neyman in 1923.

15.2 EXAMPLE

The analysis of variance can be used to describe otherwise complex relations among variables. A dog show provides an example. A dog show is not a random sampling of the breed: it is typically limited to dogs that are adult, pure-bred, and exemplary.

A histogram of dog weights from a show is likely to be rather complicated, like the yellow-orange distribution shown in the illustrations. Suppose we wanted to predict the weight of a dog based on a certain set of characteristics of each dog.

One way to do that is to *explain* the distribution of weights by dividing the dog population into groups based on those characteristics. A successful grouping will split dogs such that (a) each group has a low variance of dog weights (meaning the group is relatively homogeneous) and (b) the mean of each group is distinct (if two groups have the same mean, then it isn't reasonable to conclude that the groups are, in fact, separate in any meaningful way).

In the illustrations to the right, groups are identified as X_1, X_2, etc. In the first illustration, the dogs are divided according to the product (interaction) of two binary groupings: young vs old, and short-haired vs long-haired (e.g., group 1 is young, short-haired dogs, group 2 is young, long-haired dogs, etc.). Since the distributions of dog weight within each of the groups has a relatively large variance, and since the means are very similar across groups, grouping dogs by these characteristics does not produce an effective way to explain the variation in dog weights: knowing which group a dog is in doesn't allow us to predict its weight much better than simply knowing the dog is in a dog show. Thus, this grouping fails to explain the variation in the overall distribution.

An attempt to explain the weight distribution by grouping dogs as *pet vs working breed* and *less athletic vs more athletic* would probably be somewhat more successful (fair fit). The heaviest show dogs are likely to be big, strong, working breeds, while breeds kept as pets tend to be smaller and thus lighter. As shown by the second illustration, the distributions have variances that are considerably smaller than in the first case, and the means are more distinguishable. However, the significant overlap of distributions, for example, means that we cannot distinguish X_1 and X_2 reliably. Grouping dogs according to a coin flip might produce distributions that look similar.

An attempt to explain weight by breed is likely to produce a very good fit. All Chihuahuas are light and all St Bernards are heavy. The difference in weights between

Setters and Pointers does not justify separate breeds. The analysis of variance provides the formal tools to justify these intuitive judgments. A common use of the method is the analysis of experimental data or the development of models. The method has some advantages over correlation: not all of the data must be numeric and one result of the method is a judgment in the confidence in an explanatory relationship.

15.3 CLASSES OF MODELS

There are three classes of models used in the analysis of variance, and these are outlined here.

15.3.1 Fixed-Effects Models

The fixed-effects model (class I) of analysis of variance applies to situations in which the experimenter applies one or more treatments to the subjects of the experiment to see whether the response variable values change. This allows the experimenter to estimate the ranges of response variable values that the treatment would generate in the population as a whole.

15.3.2 Random-Effects Models

Random-effects model (class II) is used when the treatments are not fixed. This occurs when the various factor levels are sampled from a larger population. Because the levels themselves are random variables, some assumptions and the method of contrasting the treatments (a multi-variable generalization of simple differences) differ from the fixed-effects model.

15.3.3 Mixed-Effects Models

A mixed-effects model (class III) contains experimental factors of both fixed and random-effects types, with appropriately different interpretations and analysis for the two types.

15.3.4 Example

Teaching experiments could be performed by a college or university department to find a good introductory textbook, with each text considered a treatment. The fixed-effects model would compare a list of candidate texts. The random-effects model would determine whether important differences exist among a list of randomly selected texts. The mixed-effects model would compare the (fixed) incumbent texts to randomly selected alternatives. Defining fixed and random effects has proven elusive, with multiple competing definitions.

15.4 ASSUMPTIONS

The analysis of variance has been studied from several approaches, the most common of which uses a linear model that relates the response to the treatments and blocks. Note that the model is linear in parameters but may be nonlinear across factor levels. Interpretation is easy when data is balanced across factors but much deeper understanding is needed for unbalanced data.

15.4.1 Textbook Analysis Using a Normal Distribution

The analysis of variance can be presented in terms of a linear model, which makes the following assumptions about the probability distribution of the responses:

- **Independence of Observations:** This is an assumption of the model that simplifies the statistical analysis.
- **Normality:** The distributions of the residuals are normal.
- **Equality (or "Homogeneity") of Variances, Called Homoscedasticity:** The variance of data in groups should be the same.

The separate assumptions of the textbook model imply that the errors are independently, identically, and normally distributed for fixed effects models, that is, that the errors are independent and

$$\varepsilon \sim N(0, \sigma^2).$$

15.4.2 Randomization-Based Analysis

In a randomized controlled experiment, the treatments are randomly assigned to experimental units, following the experimental protocol. This randomization is objective and declared before the experiment is carried out. The objective random-assignment is used to test the significance of the null hypothesis, following the ideas of C. S. Peirce and Ronald Fisher. This design-based analysis was discussed and developed by Francis J. Anscombe at Rothamsted Experimental Station and by Oscar Kempthorne at Iowa State University. Kempthorne and his students make an assumption of *unit treatment additivity*, which is discussed in the books of Kempthorne and David R. Cox.

Derived Linear Model

Kempthorne uses the randomization-distribution and the assumption of *unit treatment additivity* to produce a *derived linear model*, very similar to the textbook model discussed previously. The test statistics of this derived linear model are closely approximated by the test statistics of an appropriate normal linear model, according to approximation theorems and simulation studies. However, there are differences. For example, the randomization-based analysis results in a small but (strictly) negative correlation between the observations. In the randomization-based analysis, there is *no assumption* of a *normal* distribution and certainly *no assumption* of *independence*. On the contrary, *the observations are dependent*!

The randomization-based analysis has the disadvantage that its exposition involves tedious algebra and extensive time. Since the randomization-based analysis is complicated and is closely approximated by the approach using a normal linear model, most teachers emphasize the normal linear model approach. Few statisticians object to model-based analysis of balanced randomized experiments.

Statistical Models for Observational Data

However, when applied to data from non-randomized experiments or observational studies, model-based analysis lacks the warrant of randomization. For observational data, the derivation of confidence intervals must use *subjective* models, as emphasized by Ronald Fisher and his followers. In practice, the estimates of treatment-effects from observational studies generally are often inconsistent. In practice, "statistical models" and observational data are useful for suggesting hypotheses that should be treated very cautiously by the public.

15.4.3 Summary of Assumptions

The normal-model based ANOVA analysis assumes the independence, normality, and homogeneity of variances of the residuals. The randomization-based analysis assumes only the homogeneity of the variances of the residuals (as a consequence of unit-treatment additivity) and uses the randomization procedure of the experiment. Both these analyses require homoscedasticity, as an assumption for the normal-model analysis and as a consequence of randomization and additivity for the randomization-based analysis.

However, studies of processes that change variances rather than means (called dispersion effects) have been successfully conducted using ANOVA. There are *no* necessary assumptions for ANOVA in its full generality, but the F-test used for ANOVA hypothesis testing has assumptions and practical limitations which are of continuing interest.

Problems which do not satisfy the assumptions of ANOVA can often be transformed to satisfy the assumptions. The property of unit-treatment additivity is not invariant under a "change of scale", so statisticians often use transformations to achieve unit-treatment additivity. If the response variable is expected to follow a parametric family of probability distributions, then the statistician may specify (in the protocol for the experiment or observational study) that the responses be transformed to stabilize the variance. Also, a statistician may specify that logarithmic transforms be applied to the responses which are believed to follow a multiplicative model. According to Cauchy's functional equation theorem, the logarithm is the only continuous transformation that transforms real multiplication to addition.

15.5 CHARACTERISTICS

ANOVA is used in the analysis of comparative experiments, those in which only the difference in outcomes is of interest. The statistical significance of the experiment is determined by a ratio of two variances. This ratio is independent of several possible alterations to the experimental observations: Adding a constant to all observations does not alter significance. Multiplying all observations by a constant does not alter significance. So ANOVA statistical significance result is independent of constant bias and scaling errors as well as the units used in expressing observations. In the era of mechanical calculation it was common to subtract a constant from all observations (when equivalent to dropping leading digits) to simplify data entry. This is an example of data coding.

15.6 ALGORITHM

The calculations of ANOVA can be characterized as computing a number of means and variances, dividing two variances and comparing the ratio to a handbook value to determine statistical significance. Calculating a treatment effect is then trivial: "the effect of any treatment is estimated by taking the difference between the mean of the observations which receive the treatment and the general mean".

15.6.1 Partitioning of the Sum of Squares

ANOVA uses traditional standardized terminology. The definitional equation of sample variance is

$$s^2 = \frac{1}{n-1}\sum_i (y_i - \overline{y})^2$$

where the divisor is called the degrees of freedom (DF), the summation is called the sum of squares (SS), the result is called the mean square (MS) and the squared terms are deviations from the sample mean. ANOVA estimates 3 sample variances: a total variance based on all the observation deviations from the grand mean, an error variance based on all the observation deviations from their appropriate treatment means, and a treatment variance. The treatment variance is based on the deviations of treatment means from the grand mean, the result being multiplied by the number of observations in each treatment to account for the difference between the variance of observations and the variance of means.

The fundamental technique is a partitioning of the total sum of squares *SS* into components related to the effects used in the model. For example, the model for a simplified ANOVA with one type of treatment at different levels.

$$SS_{\text{Total}} = SS_{\text{Error}} + SS_{\text{Treatments}}$$

The number of degrees of freedom *DF* can be partitioned in a similar way: one of these components (that for error) specifies a chi-squared distribution which describes the associated sum of squares, while the same is true for "treatments" if there is no treatment effect.

$$DF_{\text{Total}} = DF_{\text{Error}} + DF_{\text{Treatments}}$$

15.6.2 The F-Test

The *F*-test is used for comparing the factors of the total deviation. For example, in one-way, or single-factor ANOVA, statistical significance is tested for by comparing the F test statistic

$$F = \frac{\text{variance between treatments}}{\text{variance within treatments}}$$

$$F = \frac{MS_{\text{Treatments}}}{MS_{\text{Error}}} = \frac{SS_{\text{Treatments}}/(I-1)}{SS_{\text{Error}}/(n_T-I)}$$

to the *F*-distribution with $I-1$ being the numerator degrees of freedom and $n_T - I$ the denominator degrees of freedom. Using the *F*-distribution is a natural candidate because the test statistic is the ratio of two scaled sums of squares each of which follows a scaled chi-squared distribution.

The expected value of F is $1 + n\sigma^2_{\text{Treatment}} / \sigma^2_{\text{Error}}$ (where n n is the treatment sample size) which is 1 for no treatment effect. As values of F increase above 1, the evidence is increasingly inconsistent with the null hypothesis. Two apparent experimental methods of increasing F are increasing the sample size and reducing the error variance by tight experimental controls.

There are two methods of concluding the ANOVA hypothesis test, both of which produce the same result:

- The textbook method is to compare the observed value of F with the critical value of F determined from tables. The critical value of F is a function of the degrees of freedom of the numerator and the denominator and the significance level (α). If $F \geq F_{\text{Critical}}$, the null hypothesis is rejected.
- The computer method calculates the probability (p-value) of a value of F greater than or equal to the observed value. The null hypothesis is rejected if this probability is less than or equal to the significance level (α).

The ANOVA *F*-test is known to be nearly optimal in the sense of minimizing false negative errors for a fixed rate of false positive errors (i.e. maximizing power for a fixed significance level). For example, to test the hypothesis that various medical treatments have exactly the same effect, the *F*-test's *p*-values closely approximate the permutation test's p-values: The approximation is particularly close when the design is balanced. Such permutation tests characterize tests with maximum power against all alternative hypotheses, as observed by Rosenbaum. The ANOVA *F*-test (of the null-hypothesis that all treatments have exactly the same effect) is recommended as a practical test, because of its robustness against many alternative distributions.

15.6.3 Extended Algorithm

ANOVA consists of separable parts; partitioning sources of variance and hypothesis testing can be used individually. ANOVA is used to support other statistical tools. Regression is first used to fit more complex models to data, then ANOVA is used to compare models with the objective of selecting simple(r) models that adequately describe the data. "Such models could be fit without any reference to ANOVA, but ANOVA tools could then be used to make some sense of the fitted models, and to test hypotheses about batches of coefficients." "[W]e think of the analysis of variance as a way of understanding and structuring multilevel models—not as an alternative to regression but as a tool for summarizing complex high-dimensional inferences ..."

15.7 FOR A SINGLE FACTOR

The simplest experiment suitable for ANOVA analysis is the completely randomized experiment with a single factor.

More complex experiments with a single factor involve constraints on randomization and include completely randomized blocks and Latin squares (and variants: Graeco-Latin squares, etc.). The more complex experiments share many of the complexities of multiple factors.

There are some alternatives to conventional one-way analysis of variance, e.g.: Welch's heteroscedastic F test, Welch's heteroscedastic F test with trimmed means and Winsorized variances, Brown-Forsythe test, Alexander-Govern test, James second order test and Kruskal-Wallis test, available in onewaytests R

It is useful to represent each data point in the following form, called a statistical model:

$$Y_{ij} = \mu + \tau_j + \varepsilon_{ij}$$

where

- $i = 1, 2, 3, ..., R$
- $j = 1, 2, 3, ..., C$
- μ = overall average (mean)
- τ_j = differential effect (response) associated with the j level of X; this assumes that overall the values of τ_j add to zero (that is $\sum_{j=1}^{C} \tau_j = 0$)
- ε_{ij} = noise or error associated with the particular ij data value

That is, we envision an additive model that says every data point can be represented by summing three quantities: the true mean, averaged over all factor levels being investigated, plus an incremental component associated with the particular column (factor level), plus a final component associated with everything else affecting that specific data value.

15.8 FOR MULTIPLE FACTORS

ANOVA generalizes to the study of the effects of multiple factors. When the experiment includes observations at all combinations of levels of each factor, it is termed factorial. Factorial experiments are more efficient than a series of single factor experiments and the efficiency grows as the number of factors increases. Consequently, factorial designs are heavily used.

The use of ANOVA to study the effects of multiple factors has a complication. In a 3-way ANOVA with factors x, y and z, the ANOVA model includes terms for the main effects (x, y, z) and terms for interactions (xy, xz, yz, xyz). All terms require hypothesis tests. The proliferation of interaction terms increases the risk that some hypothesis test will produce a false positive by chance. Fortunately, experience says that high order interactions are rare. The ability to detect interactions is a major advantage of multiple factor ANOVA. Testing one factor at a time hides interactions, but produces apparently inconsistent experimental results.

Caution is advised when encountering interactions; Test interaction terms first and expand the analysis beyond ANOVA if interactions are found. Texts vary in their recommendations regarding the continuation of the ANOVA procedure after encountering an interaction. Interactions complicate the interpretation of experimental data. Neither the calculations of significance nor the estimated treatment effects can be taken at face value. "A significant interaction will often mask the significance of main effects." Graphical methods are recommended to enhance understanding. Regression is often useful. A lengthy discussion of interactions is available in Cox (1958). Some interactions can be removed (by transformations) while others cannot.

A variety of techniques are used with multiple factor ANOVA to reduce expense. One technique used in factorial designs is to minimize replication (possibly no replication with support of analytical trickery) and to combine groups when effects are found to be statistically (or practically) insignificant. An experiment with many insignificant factors may collapse into one with a few factors supported by many replications.

15.9 ASSOCIATED ANALYSIS

Some analysis is required in support of the *design* of the experiment while other analysis is performed after changes in the factors are formally found to produce statistically significant changes in the responses. Because experimentation is iterative, the results of one experiment alter plans for following experiments.

15.9.1 Preparatory Analysis

The Number of Experimental Units

In the design of an experiment, the number of experimental units is planned to satisfy the goals of the experiment. Experimentation is often sequential. Early experiments are often designed to provide mean-unbiased estimates of treatment effects and of experimental error.

Later experiments are often designed to test a hypothesis that a treatment effect has an important magnitude; in this case, the number of experimental units is chosen so that the experiment is within budget and has adequate power, among other goals.

Reporting sample size analysis is generally required in psychology. "Provide information on sample size and the process that led to sample size decisions." The analysis, which is written in the experimental protocol before the experiment is conducted, is examined in grant applications and administrative review boards.

Besides the power analysis, there are less formal methods for selecting the number of experimental units. These include graphical methods based on limiting the probability of false negative errors, graphical methods based on an expected variation increase (above the residuals) and methods based on achieving a desired confidence interval.

Power Analysis

Power analysis is often applied in the context of ANOVA in order to assess the probability of successfully rejecting the null hypothesis if we assume a certain ANOVA design, effect size in the population, sample size and significance level. Power analysis can assist in study design by determining what sample size would be required in order to have a reasonable chance of rejecting the null hypothesis when the alternative hypothesis is true.

Effect Size

Several standardized measures of effect have been proposed for ANOVA to summarize the strength of the association between a predictor(s) and the dependent variable or the overall standardized difference of the complete model. Standardized effect-size estimates facilitate comparison of findings across studies and disciplines. However, while standardized effect sizes are commonly used in much of the professional literature, a non-standardized measure of effect size that has immediately "meaningful" units may be preferable for reporting purposes.

Model Confirmation

Sometimes tests are conducted to determine whether the assumptions of ANOVA appear to be violated. Residuals are examined or analyzed to confirm homoscedasticity and gross normality. Residuals should have the appearance of (zero mean normal distribution) noise when plotted as a function of anything including time and modeled data values. Trends hint at interactions among factors or among observations.

Follow-Up Tests

A statistically significant effect in ANOVA is often followed by additional tests. This can be done in order to assess which groups are different from which other groups or to test various other focused hypotheses. Follow-up tests are often distinguished in terms of whether they are "planned" (a priori) or "post hoc." Planned tests are determined before looking at the data, and post hoc tests are conceived only after looking at the data (though the term "post hoc" is inconsistently used).

The follow-up tests may be "simple" pairwise comparisons of individual group means or may be "compound" comparisons (e.g., comparing the mean pooling across groups A, B and C to the mean of group D). Comparisons can also look at tests of trend, such as linear and quadratic relationships, when the independent variable involves ordered levels. Often the follow-up tests incorporate a method of adjusting for the multiple comparisons problem.

Follow-up tests to identify which specific groups, variables, or factors have statistically different means include the Tukey's range test, and Duncan's new multiple range test. In turn, these tests are often followed with a Compact Letter Display (CLD) methodology in order to render the output of the mentioned tests more transparent to a non-statistician audience.

15.10 STUDY DESIGNS

There are several types of ANOVA. Many statisticians base ANOVA on the design of the experiment, especially on the protocol that specifies the random assignment of treatments to subjects; the protocol's description of the assignment mechanism should include a specification of the structure of the treatments and of any blocking. It is also common to apply ANOVA to observational data using an appropriate statistical model.

Some popular designs use the following types of ANOVA:

- One-way ANOVA is used to test for differences among two or more independent groups (means), e.g. different levels of urea application in a crop, or different levels of antibiotic action on several different bacterial species, or different levels of effect of some medicine on groups of patients. However, should these groups not be independent, and there is an order in the groups (such as mild, moderate and severe disease), or in the dose of a drug (such as 5 mg/mL, 10 mg/mL, 20 mg/mL) given to the same group of patients, then a linear trend estimation should be used. Typically, however, the one-way ANOVA is used to test for differences among at least three groups, since the two-group case can be covered by a t-test. When there are only two means to compare, the t-test and the ANOVA *F*-test are equivalent; the relation between ANOVA and *t* is given by $F = t^2$.

- Factorial ANOVA is used when there is more than one factor.
- Repeated measures ANOVA is used when the same subjects are used for each factor (e.g., in a longitudinal study).
- Multivariate analysis of variance (MANOVA) is used when there is more than one response variable.

15.11 CAUTIONS

Balanced experiments (those with an equal sample size for each treatment) are relatively easy to interpret; unbalanced experiments offer more complexity. For single-factor (one-way) ANOVA, the adjustment for unbalanced data is easy, but the unbalanced analysis lacks both robustness and power. For more complex designs the lack of balance leads to further complications. "The orthogonality property of main effects and interactions present in balanced data does not carry over to the unbalanced case. This means that the usual analysis of variance techniques do not apply. Consequently, the analysis of unbalanced factorials is much more difficult than that for balanced designs." In the general case, "The analysis of variance can also be applied to unbalanced data, but then the sums of squares, mean squares, and *F*-ratios will depend on the order in which the sources of variation are considered."

ANOVA is (in part) a test of statistical significance. The American Psychological Association (and many other organisations) holds the view that simply reporting statistical significance is insufficient and that reporting confidence bounds is preferred.

CHAPTER 16

NON-PARAMETRIC OR DISTRIBUTION-FREE STATISTICAL TESTS

Non-parametric or distribution-free statistical tests are a class of statistical methods that do not require the assumption of a specific probability distribution for the data. These tests are particularly useful when the assumptions of parametric tests (such as normality) cannot be met or when the data is ordinal, categorical, or skewed. Unlike parametric tests, which rely on parameters like the mean and standard deviation, non-parametric tests focus on the order or rank of data, making them more flexible in handling a wide variety of data types.

Non-parametric tests play a crucial role in various fields such as medicine, biology, social sciences, and finance, where data often deviates from normality or does not meet the stringent requirements of parametric tests.

16.1 WHY USE NON-PARAMETRIC TESTS?

Non-parametric tests are used in situations where the data does not satisfy the assumptions required for parametric tests like the t-test or ANOVA. Some of the key situations in which non-parametric tests are appropriate include:

- **Non-Normal Data:** Parametric tests assume that the data follows a normal distribution. Non-parametric tests are useful when the data is skewed, has outliers, or is otherwise non-normally distributed.
- **Ordinal Data:** Non-parametric tests are suited for ordinal data, where the exact differences between ranks are not known but the relative ranking is meaningful. Parametric tests typically require interval or ratio data, where the distance between data points is measurable.
- **Small Sample Sizes:** When the sample size is small, parametric tests may not be robust enough, particularly in terms of meeting the normality assumption. Non-parametric tests do not rely on this assumption and are often more appropriate for smaller datasets.
- **Robustness to Outliers:** Non-parametric tests are less sensitive to outliers, which can significantly distort the results of parametric tests.

16.2 ADVANTAGES OF NON-PARAMETRIC TESTS

There are several advantages to using non-parametric tests:

- **Fewer Assumptions:** Non-parametric tests do not require assumptions about the underlying population distribution, making them more flexible and broadly applicable to different types of data.
- **Handling of Small Samples:** Non-parametric tests are ideal for small sample sizes, where parametric methods might fail due to unreliable estimates of the population parameters.
- **Ordinal and Categorical Data:** These tests are designed to handle ordinal and categorical data, which are often encountered in fields like psychology, sociology, and market research.
- **Robustness:** Non-parametric tests are less sensitive to outliers and extreme values, ensuring that the results are more reliable when the data contains unusual values.

16.3 DISADVANTAGES OF NON-PARAMETRIC TESTS

Despite their flexibility, non-parametric tests have some limitations:

- **Less Power:** Non-parametric tests are generally less powerful than parametric tests when the assumptions of parametric tests are met. This means that non-parametric tests may require larger sample sizes to detect the same effect.
- **Limited Interpretation:** While parametric tests provide estimates of parameters such as means and variances, non-parametric tests do not provide such detailed information. Instead, they focus on ranks or medians, which limits their interpretative power.
- **Data Requirements:** Non-parametric tests often require that the data be at least ordinal, meaning that the data can be ranked. For certain types of data, this may not be possible.

16.4 COMMON NON-PARAMETRIC TESTS

Several widely-used non-parametric tests cater to different types of data and research questions. Below are some of the most common non-parametric tests:

16.4.1 Wilcoxon Rank-Sum Test (Mann-Whitney U Test)

The **Wilcoxon rank-sum test**, also known as the **Mann-Whitney U test**, is a non-parametric alternative to the independent samples t-test. It is used to compare two independent groups to determine if one group tends to have larger values than the other. Instead of comparing means, the test compares the ranks of the values in the two groups.

This test is especially useful when the data is ordinal, not normally distributed, or when there are outliers that could distort the results of a t-test. The test works by ranking all the data points across both groups and then comparing the sum of ranks between the groups.

Example:

- **Research Question:** Does a new medication reduce pain more effectively than a placebo in a group of patients?
- **Data Type:** Ordinal pain scores (e.g., "no pain," "mild pain," "moderate pain," "severe pain").

16.4.2 Wilcoxon Signed-Rank Test

The **Wilcoxon signed-rank test** is a non-parametric test used to compare two related samples or paired observations. It is the non-parametric alternative to the paired t-test. The test evaluates whether there is a difference in the median ranks of the paired observations. This test is used when the data is not normally distributed but consists of paired samples, such as pre-test and post-test scores for the same individuals.

Example:

- **Research Question:** Is there a difference in blood pressure before and after administering a treatment in the same group of patients?
- **Data Type:** Paired, ordinal data (e.g., blood pressure measurements).

16.4.3 Kruskal-Wallis Test

The **Kruskal-Wallis test** is the non-parametric counterpart to one-way ANOVA. It is used to compare the medians of three or more independent groups to determine if they come from the same population. This test is based on the ranks of the data rather than the actual data values. Like other non-parametric tests, the Kruskal-Wallis test is ideal when the assumptions of ANOVA (such as normality and homogeneity of variances) are violated.

Example:

- **Research Question:** Are there differences in patient satisfaction scores across three different hospital departments?
- **Data Type:** Ordinal satisfaction scores.

16.4.4 Chi-Square Test

The **chi-square test** is a non-parametric test used to examine the association between two categorical variables. It is often applied in cross-tabulation or contingency table analysis to determine whether the distribution of one variable differs across the levels of another variable. This test compares the observed frequencies in the data to the frequencies that would be expected if the two variables were independent.

Example:

- **Research Question:** Is there an association between gender and preference for a particular brand of product?
- **Data Type:** Categorical data (e.g., gender: male/female, preference: brand A/brand B).

16.4.5 Spearman's Rank Correlation Coefficient

Spearman's rank correlation is a non-parametric measure of the strength and direction of the association between two ranked variables. It is used when the data is ordinal or when the relationship between variables is monotonic but not necessarily linear.

Spearman's rank correlation is less sensitive to outliers than Pearson's correlation coefficient.

Example:

- **Research Question:** Is there a relationship between students' rank in their class and their performance on a standardized test?
- **Data Type:** Ordinal rankings.

16.4.6 Friedman Test

The **Friedman test** is a non-parametric test used to compare three or more related groups. It is the non-parametric alternative to repeated measures ANOVA.

The test evaluates whether the ranks of the data differ across multiple conditions or time points for the same group of subjects.

Example:

- **Research Question:** Do reaction times differ when subjects perform a task under three different lighting conditions?
- **Data Type:** Repeated measures of ordinal data (e.g., reaction times across conditions).

16.5 CONCLUSION

Non-parametric tests provide a flexible and robust alternative to parametric tests when the assumptions of the latter cannot be met.

They are invaluable in dealing with small sample sizes, non-normal data, ordinal data, and outliers. While non-parametric tests may not be as powerful as parametric tests when parametric assumptions hold, they offer a reliable way to analyze data in many real-world scenarios where data do not conform to the idealized conditions of parametric methods.

By using ranks or categorical data rather than relying on specific distributions, non-parametric tests broaden the range of statistical tools available to researchers, enabling them to make meaningful inferences even in the face of data challenges. As data analysis continues to evolve across disciplines, the role of non-parametric tests in providing robust, flexible solutions to complex problems will remain critical.

CHAPTER 17
VITAL STATISTICS

Vital statistics refer to the systematic collection, analysis, and interpretation of data related to important life events in a population, such as births, deaths, marriages, and divorces. These statistics are crucial for understanding the health and demographic profile of a population and are used by governments, researchers, and public health officials to inform policy decisions and improve public health outcomes. Vital statistics provide essential data for understanding population growth, life expectancy, mortality rates, and the overall health of a community.

Vital statistics play a key role in a wide range of areas, including public health planning, resource allocation, monitoring and evaluation of health programs, and understanding trends in disease outbreaks or other health-related events.

17.1 COMPONENTS OF VITAL STATISTICS

17.1.1 Birth Statistics

Birth statistics are one of the most fundamental components of vital statistics. They provide information on the number of live births in a population, as well as various demographic factors related to childbirth, such as maternal age, marital status, and place of birth. Birth statistics help public health officials understand trends in fertility rates and reproductive health. Birth data are important for tracking population growth, planning for educational and healthcare needs, and understanding maternal and child health trends. By analyzing birth statistics, public health officials can monitor changes in fertility patterns, identify areas with high rates of teenage pregnancies, and evaluate the impact of programs aimed at improving maternal and infant health.

17.1.2 Death Statistics

Death statistics provide information about mortality in a population, including the number of deaths, causes of death, and demographic characteristics such as age, gender, and geographic location. Mortality data are essential for understanding life expectancy, identifying leading causes of death, and tracking health disparities within different population groups. Death statistics are crucial for monitoring the overall health of a population and identifying public health challenges. For example, high mortality rates from diseases such as heart disease, cancer, or infectious diseases may indicate the need for targeted public health interventions.

Similarly, mortality data can help track the impact of emerging health threats, such as pandemics or natural disasters.

17.1.3 Marriage and Divorce Statistics

Marriage and divorce statistics provide valuable information on family dynamics and social trends. These statistics help track changes in marriage and divorce rates over time, as well as factors related to family formation and dissolution. Researchers and policymakers use these data to understand social changes, such as shifts in family structures, marriage trends, and the impact of divorce on children and society. Marriage and divorce statistics are important for understanding the broader social and cultural context in which people live. For example, rising divorce rates may highlight the need for support services for families, while changes in marriage patterns can provide insight into shifts in social norms and economic conditions.

17.1.4 Fetal Death and Stillbirth Statistics

Fetal death and stillbirth statistics refer to the death of a fetus after 20 weeks of gestation or after the fetus has reached a certain weight threshold. These statistics are crucial for monitoring maternal and fetal health and identifying factors that contribute to poor pregnancy outcomes. High rates of fetal deaths may indicate underlying health issues or social determinants of health, such as access to prenatal care or environmental hazards. Fetal death data are used by healthcare providers and public health officials to design and implement programs aimed at reducing stillbirths and improving prenatal care. By analyzing these data, researchers can identify risk factors and develop interventions to prevent fetal deaths.

17.2 METHODS OF DATA COLLECTION

17.2.1 Civil Registration Systems

Civil registration systems are the primary source of vital statistics in most countries. These systems record vital events such as births, deaths, marriages, and divorces through a legal framework that mandates registration. Civil registration systems ensure that vital events are recorded accurately and comprehensively, providing a continuous and reliable source of data for governments and researchers. Civil registration systems are often supplemented by other methods of data collection, such as household surveys and censuses, particularly in countries where civil registration systems are incomplete or underdeveloped.

17.2.2 Household Surveys

In some cases, especially in countries where civil registration systems are not fully established, household surveys may be used to collect vital statistics. These surveys gather data directly from households through questionnaires or interviews. Household surveys can provide important information about demographic and health trends, but they may be subject to limitations such as recall bias or incomplete responses. Household surveys are particularly useful for collecting data in regions with low rates of birth and death registration, allowing public health officials to fill gaps in civil registration data.

17.2.3 Censuses

Censuses are another method for collecting vital statistics, although they are conducted less frequently than civil registration or household surveys. A census collects data on a wide range of demographic and social factors, including vital events such as births and deaths. While censuses provide valuable data, they are typically conducted only once every 10 years, limiting their usefulness for tracking short-term trends in vital statistics.

17.3 IMPORTANCE OF ACCURATE AND TIMELY DATA

The accuracy and timeliness of vital statistics are critical for their usefulness in public health planning and policy-making. Reliable data enable governments and health organizations to make informed decisions about resource allocation, program development, and the implementation of public health interventions. Timely data are essential for tracking health trends and responding to emerging public health threats. For example, during a disease outbreak, up-to-date mortality data can help public health officials identify areas where the outbreak is most severe and allocate resources accordingly. Delayed or inaccurate data can hinder efforts to control disease outbreaks, monitor health trends, and assess the impact of public health interventions.

17.4 CHALLENGES IN COLLECTING VITAL STATISTICS

17.4.1 Inadequate Civil Registration Systems

In many low- and middle-income countries, civil registration systems are incomplete or underdeveloped, leading to gaps in vital statistics. In some cases, births and deaths may go unregistered, particularly in rural or remote areas where access to registration offices is limited. Inadequate civil registration systems can lead to inaccurate estimates of population size, fertility rates, and mortality rates, which can in turn affect public health planning and resource allocation.

17.4.2 Data Quality Issues

Even in countries with well-established civil registration systems, data quality issues can arise. These may include incomplete or inaccurate reporting of vital events, particularly in cases where healthcare providers or registrars lack training in data collection procedures. Ensuring the accuracy of vital statistics requires ongoing efforts to improve data collection methods, train registrars, and enforce legal requirements for registration.

17.4.3 Political and Cultural Barriers

In some countries, cultural or political barriers may prevent the accurate collection of vital statistics. For example, cultural norms surrounding birth registration may lead to delays in registering newborns, particularly in countries where traditional practices dictate the timing of registration. Political instability or conflict may also disrupt civil registration systems, making it difficult to collect and analyze vital statistics.

17.4.4 Technological Limitations

Technological limitations can also pose challenges to the collection of vital statistics. In some regions, particularly in low-income countries, there may be limited access to the technology needed to collect, store, and analyze data. This can result in delays in data processing and reporting, as well as challenges in sharing data with public health authorities.

17.5 CONCLUSION

Vital statistics are an essential tool for understanding the health and demographic profile of a population. Accurate and timely data on births, deaths, marriages, and other vital events provide critical insights into public health trends, social dynamics, and population growth. These statistics play a central role in public health planning, resource allocation, and the development of targeted interventions to improve health outcomes.

Despite their importance, the collection and analysis of vital statistics face numerous challenges, including inadequate civil registration systems, data quality issues, and technological limitations. Addressing these challenges is critical for ensuring that vital statistics remain a reliable source of data for policymakers, healthcare providers, and researchers. By improving data collection methods and investing in civil registration systems, governments can ensure that vital statistics continue to play a key role in shaping public health policy and improving population health.

CHAPTER 18
HEALTH STATISTICS

Health statistics are an essential tool for understanding and improving public health. They encompass data related to health outcomes, healthcare utilization, disease prevalence, and mortality rates. Health statistics provide a quantitative foundation for making informed decisions about healthcare policy, planning, and intervention strategies. They are crucial in identifying trends, monitoring progress toward health goals, and understanding disparities in health outcomes across different population groups.

Health statistics serve multiple stakeholders, including government agencies, healthcare providers, researchers, and the general public. By analyzing this data, public health professionals can better understand the causes of illness and death, identify populations at risk, and evaluate the effectiveness of healthcare systems.

18.1 COMPONENTS OF HEALTH STATISTICS

18.1.1 Morbidity Statistics

Morbidity statistics refer to data on the incidence and prevalence of diseases and other health conditions within a population. Incidence measures the number of new cases of a disease over a specific time period, while prevalence measures the total number of existing cases at a particular point in time. Morbidity statistics are vital for understanding the burden of disease in a community and for designing interventions to prevent and treat illness. Morbidity data help public health officials monitor the spread of infectious diseases, track chronic health conditions such as diabetes and heart disease, and identify emerging health threats. This information can be used to allocate resources effectively and to develop health education and prevention programs aimed at reducing the incidence of specific diseases.

18.1.2 Mortality Statistics

Mortality statistics, which refer to data on the number of deaths in a population, provide critical information about the overall health of a community. These statistics include death rates, causes of death, and demographic factors such as age, sex, and geographic location. Mortality data are essential for tracking changes in life expectancy and identifying leading causes of death. Mortality statistics allow public health professionals to identify trends in premature deaths, evaluate the impact of public health interventions, and monitor the effects of pandemics or other health crises. For example, during the COVID-19 pandemic, mortality statistics were crucial in understanding the scale and progression of the virus and in determining the effectiveness of interventions such as vaccines and social distancing measures.

18.1.3 Healthcare Utilization Statistics

Healthcare utilization statistics provide data on how often and why people use healthcare services, such as hospital admissions, outpatient visits, emergency room usage, and prescription drug consumption. These statistics help identify patterns in healthcare access and utilization, as well as potential barriers to care. By analyzing healthcare utilization data, policymakers can assess the efficiency and effectiveness of healthcare systems, identify gaps in service delivery, and plan for future healthcare needs. For example, healthcare utilization statistics may reveal disparities in access to care based on factors such as income, race, or geographic location, allowing for targeted interventions to reduce these disparities.

18.1.4 Health Behavior Statistics

Health behavior statistics provide information on behaviors that influence health outcomes, such as smoking, alcohol consumption, physical activity, and dietary habits. These behaviors are closely linked to the risk of developing chronic diseases such as cancer, heart disease, and diabetes. Public health professionals use health behavior statistics to design and evaluate programs aimed at promoting healthy lifestyles and reducing risky behaviors. For example, smoking cessation programs or initiatives to promote physical activity can be tailored based on data showing the prevalence of these behaviors in specific populations.

18.1.5 Health System Performance Statistics

Health system performance statistics evaluate the quality, efficiency, and equity of healthcare delivery. These statistics may include data on patient satisfaction, treatment outcomes, hospital readmission rates, and healthcare costs. Health system performance data help healthcare administrators and policymakers assess how well healthcare services meet the needs of the population. By analyzing health system performance statistics, decision-makers can identify areas where healthcare systems need improvement, such as reducing medical errors, improving patient safety, or increasing access to preventive care services.

18.2 METHODS OF DATA COLLECTION

18.2.1 Surveys and Questionnaires

Surveys and questionnaires are widely used to collect health data from individuals or households. These instruments can gather information on a wide range of topics, including health behaviors, healthcare utilization, and self-reported health status. Examples of major health surveys include the National Health and Nutrition Examination Survey (NHANES) and the Behavioral Risk Factor Surveillance System (BRFSS). Surveys provide valuable data on health trends and allow researchers to gather information from a broad and diverse population. However, surveys also have limitations, such as potential recall bias, where respondents may not accurately remember or report their health behaviors or medical history.

18.2.2 Vital Records

Vital records, such as birth certificates, death certificates, and marriage licenses, are an essential source of health statistics. These records provide critical information on life events and demographic characteristics. Birth and death certificates, for example, contain data on maternal and infant health, causes of death, and life expectancy. Vital records are considered highly reliable and are used by public health authorities to track important health trends, such as infant mortality rates, life expectancy, and the leading causes of death. These records are typically collected by government agencies through civil registration systems.

18.2.3 Electronic Health Records (EHRs)

Electronic Health Records (EHRs) are digital systems that store patient medical information, including diagnoses, treatments, medications, and laboratory results. EHRs are increasingly being used as a source of health statistics because they provide real-time data on healthcare delivery and patient outcomes. EHRs allow healthcare providers and researchers to track patient care over time, monitor treatment outcomes, and identify patterns in healthcare utilization. They also facilitate data sharing between healthcare providers, improving care coordination and patient outcomes.

18.2.4 Disease Registries

Disease registries are databases that collect and store information on individuals diagnosed with specific diseases, such as cancer, diabetes, or HIV/AIDS. These registries provide a comprehensive source of data on disease incidence, treatment outcomes, and survival rates. Disease registries are particularly valuable for monitoring chronic diseases and evaluating the effectiveness of treatment protocols. For example, cancer registries allow researchers to track the progression of cancer cases, analyze survival rates, and assess the impact of new treatments.

18.3 ROLE OF HEALTH STATISTICS IN PUBLIC HEALTH

18.3.1 Planning and Policy Development

Health statistics play a crucial role in public health planning and policy development. Policymakers rely on accurate and up-to-date health data to allocate resources, set priorities, and design programs aimed at improving population health. For example, data on childhood vaccination rates can guide immunization campaigns, while statistics on obesity can inform initiatives to promote healthy eating and physical activity.

By analyzing health statistics, public health officials can identify emerging health threats, such as infectious disease outbreaks or rising rates of chronic diseases, and take proactive measures to address these challenges. Health statistics also provide the evidence base for evaluating the effectiveness of public health interventions and policies.

18.3.2 Monitoring Health Disparities

Health statistics are essential for monitoring health disparities across different population groups. Disparities in health outcomes can arise due to factors such as income, race, ethnicity, gender, or geographic location. By analyzing health data, public health officials can identify populations at risk and develop targeted interventions to reduce health inequalities. For example, mortality statistics may reveal higher death rates from certain diseases in low-income or minority populations, prompting efforts to improve access to healthcare and address social determinants of health. Similarly, healthcare utilization statistics can highlight barriers to care, such as lack of insurance or transportation, that disproportionately affect vulnerable populations.

18.3.3 Evaluating Public Health Programs

Health statistics are used to evaluate the effectiveness of public health programs and interventions. By collecting and analyzing data on health outcomes, public health professionals can determine whether a program is achieving its goals, identify areas for improvement, and make data-driven decisions about resource allocation. For example, a smoking cessation program can be evaluated by tracking smoking rates before and after the intervention. Similarly, vaccination programs can be assessed by analyzing immunization coverage rates and the incidence of vaccine-preventable diseases.

18.4 CHALLENGES IN COLLECTING AND INTERPRETING HEALTH STATISTICS

18.4.1 Data Quality and Accuracy

Ensuring the accuracy and quality of health statistics is a major challenge. Errors in data collection, reporting, or analysis can lead to inaccurate conclusions and misguided policy decisions. For example, underreporting of disease cases or deaths may result in an incomplete understanding of public health trends. Improving data quality requires investments in training for data collectors, standardizing data collection methods, and implementing quality control measures to verify the accuracy of reported information.

18.4.2 Privacy and Confidentiality Concerns

The collection of health data raises important privacy and confidentiality concerns. Protecting patient information is critical to maintaining public trust and ensuring that individuals feel comfortable providing accurate data. Governments and healthcare organizations must implement strong data protection measures to safeguard sensitive health information and comply with privacy regulations, such as the Health Insurance Portability and Accountability Act (HIPAA) in the United States.

18.4.3 Data Integration and Standardization

Health data are often collected from a variety of sources, including surveys, EHRs, and disease registries. Integrating and standardizing these data sources can be challenging due to differences in data formats, collection methods, and definitions. For example, different healthcare providers may use different coding systems for diagnoses or treatments, making it difficult to compare data across institutions. Efforts to standardize data collection and reporting methods, such as the adoption of common coding systems or data-sharing platforms, can help overcome these challenges and improve the utility of health statistics.

18.5 CONCLUSION

Health statistics are a vital tool for understanding population health, identifying public health challenges, and guiding policy decisions. From morbidity and mortality data to healthcare utilization and health behavior statistics, these data provide the foundation for effective public health planning, monitoring, and evaluation. However, collecting and interpreting health statistics is not without challenges, including issues related to data quality, privacy, and standardization. By addressing these challenges and investing in robust data collection and analysis systems, health statistics can continue to play a critical role in improving health outcomes and reducing health disparities.

CHAPTER 19

STANDARDIZED DEATH RATES AND LIFE TABLES

19.1 DEATH RATE

Mortality rate, or death rate, is a measure of the number of deaths (in general, or due to a specific cause) in a particular population, scaled to the size of that population, per unit of time. Mortality rate is typically expressed in units of deaths per 1,000 individuals per year; thus, a mortality rate of 9.5 (out of 1,000) in a population of 1,000 would mean 9.5 deaths per year in that entire population, or 0.95% out of the total. It is distinct from "morbidity", which is either the prevalence or incidence of a disease, and also from the incidence rate (the number of newly appearing cases of the disease per unit of time).

19.1.1 Crude Death Rate, Globally

According to the World Health Organization, the ten leading causes of death, globally, in 2016, for both sexes and all ages, were as presented in the table below.

Table 1: Crude Death Rate, Per 100,000 Population

Ischaemic heart disease	126
Stroke	77
Chronic obstructive pulmonary disease	41
Lower respiratory infections	40
Alzheimer's disease and other dementias	27
Trachea, bronchus, and lung cancers	23
Diabetes mellitus	21
Road injury	19
Diarrhoeal diseases	19
Tuberculosis	17

The crude death rate is defined as "the mortality rate from all causes of death for a population," calculated as the "total number of deaths during a given time interval" divided by the "mid-interval population", per 1,000 or 100,000; for instance, the population of the U.S. was around 290,810,000 in 2003, and in that year, approximately 2,419,900 deaths occurred in total, giving a crude death (mortality) rate of 832 deaths per 100,000. As of 2020, the CIA estimates the U.S. crude death rate will be 8.3 per 1,000, while it estimates that the global rate will be 7.7 per 1,000.

Mortality rate is also measured per thousand. It is determined by how many people of a certain age die per thousand people. Decrease of mortality rate is one of the reasons for increase of population. Development of medical science and other technologies has resulted in the decrease of mortality rate in all the countries of the world for some decades. In 1990, the mortality rate of children under five years of age was 144 per thousand, but in 2015 the child mortality rate was 38 per thousand.

19.1.2 Related Measures of Mortality

Other specific measures of mortality include:

Table 2: Measures of Mortality

Name	Typical Definition
Perinatal mortality rate	The sum of fetal deaths (stillbirths) past 22 (or 28) completed weeks of pregnancy plus the number of deaths among live-born children up to 7 completed days of life, divided by number of births.
Maternal mortality rate	Number of deaths of mothers assigned to pregnancy-related causes during a given time interval, divided by the number of live births during the same time interval.
Infant mortality rate	Number of deaths among children <one year of age during a given time interval divided by the number of live births during the same time interval.
Child mortality rate (also known as 'Under-five mortality rate')	Number of deaths of children less than 5 years old, divided by number of live births.
Standardized mortality ratio (SMR)	The ratio of the number of deaths in a given (index) population to the number of deaths expected, a form of indirectly (as opposed to directly) standardized rates, where the categories are usually "defined by age, gender and race or ethnicity". It has also been described as a proportional comparison to the numbers of deaths that would have been expected if the population had been of a standard composition in terms of age, gender, etc.
Age-specific mortality rate (ASMR)	The total number of deaths per year at a specific age, divided by the number of living persons at that age (e.g. age 62 at last birthday)
Cause-specific death rate	Number of deaths assigned to a specific cause during a given time interval divided by the mid-interval population

Name	Typical Definition
Cumulative death rate	The incidence proportion of death, that is, the proportion of a [defined] group that dies over a specified time interval, whether in reference to all deaths over the time inverval, to "to deaths from a specific cause or causes". It has also been described as a measure of the (growing) proportion of a group that die over a specified period (often as estimated by techniques that account for missing data by statistical censoring).
Case fatality rate (CFR)	The proportion of diagnosed cases of a particular medical condition that lead to death.
Infection fatality rate (IFR)	The proportion of infected cases of a particular medical condition that lead to death. Similar to CFR, but adjusted for asymptomatic and undiagnosed cases.

For any of these, a "sex-specific mortality rate" refers to "a mortality rate among either males or females", where the calculation involves both "numerator and denominator... limited to the one sex".

19.1.3 Use in Epidemiology

In most cases there are few if any ways to obtain exact mortality rates, so epidemiologists use estimation to predict correct mortality rates. Mortality rates are usually difficult to predict due to language barriers, health infrastructure related issues, conflict, and other reasons. Maternal mortality has additional challenges, especially as they pertain to stillbirths, abortions, and multiple births. In some countries, during the 1920s, a stillbirth was defined as "a birth of at least twenty weeks' gestation in which the child shows no evidence of life after complete birth". In most countries, however, a stillbirth was defined as "the birth of a fetus, after 28 weeks of pregnancy, in which pulmonary respiration does not occur".

Census Data and Vital Statistics

Ideally, all mortality estimation would be done using vital statistics and census data. Census data will give detailed information about the population at risk of death. The vital statistics provide information about live births and deaths in the population. Often, either census data and vital statistics data is not available. This is common in developing countries, countries that are in conflict, areas where natural disasters have caused mass displacement, and other areas where there is a humanitarian crisis

Household Surveys

Household surveys or interviews are another way in which mortality rates are often assessed. There are several methods to estimate mortality in different segments of the population. One such example is the sisterhood method, which involves researchers estimating maternal mortality by contacting women in populations of interest and asking whether or not they have a sister, if the sister is of childbearing age (usually 15) and conducting an interview or written questions about possible deaths among sisters. The sisterhood method, however, does not work in cases where sisters may have died before the sister being interviewed was born.

Orphanhood surveys estimate mortality by questioning children are asked about the mortality of their parents. It has often been criticized as an adult mortality rate that is very biased for several reasons. The adoption effect is one such instance in which orphans often do not realize that they are adopted. Additionally, interviewers may not realize that an adoptive or foster parent is not the child's biological parent. There is also the issue of parents being reported on by multiple children while some adults have no children, thus are not counted in mortality estimates.

Widowhood surveys estimate adult mortality by responding to questions about the deceased husband or wife. One limitation of the widowhood survey surrounds the issues of divorce, where people may be more likely to report that they are widowed in places where there is the great social stigma around being a divorcee. Another limitation is that multiple marriages introduce biased estimates, so individuals are often asked about first marriage. Biases will be significant if the association of death between spouses, such as those in countries with large AIDS epidemics.

Sampling

Sampling refers to the selection of a subset of the population of interest to efficiently gain information about the entire population. Samples should be representative of the population of interest. Cluster sampling is an approach to non-probability sampling; this is an approach in which each member of the population is assigned to a group (cluster), and then clusters are randomly selected, and all members of selected clusters are included in the sample. Often combined with stratification techniques (in which case it is called multistage sampling), cluster sampling is the approach most often used by epidemiologists. In areas of forced migration, there is more significant sampling error. Thus cluster sampling is not the ideal choice.

19.1.4 Mortality Statistics

Causes of death vary greatly between developed and less developed countries. According to Jean Ziegler (the United Nations Special Rapporteur on the Right to Food for 2000 to March 2008), mortality due to malnutrition accounted for 58% of the total mortality in 2006: "In the world, approximately 62 million people, all causes of death combined, die each year. In 2006, more than 36 million died of hunger or diseases due to deficiencies in micronutrients". Of the roughly 150,000 people who die each day across the globe, about two thirds—100,000 per day—die of age-related causes. In industrialized nations, the proportion is much higher, reaching 90%.

Table 3: World Historical and Predicted Crude Death Rates (1950–2050)

Years	CDR	Years	CDR
1950–1955	19.1	2000–2005	8.4
1955–1960	17.3	2005–2010	8.1
1960–1965	16.2	2010–2015	8.1
1965–1970	12.9	2015–2020	8.1
1970–1975	11.6	2020–2025	8.1
1975–1980	10.6	2025–2030	8.3
1980–1985	10.0	2030–2035	8.6
1985–1990	9.4	2035–2040	9.0
1990–1995	9.1	2040–2045	9.4
1995–2000	8.8	2045–2050	9.7

19.1.5 Economics

Scholars have stated that there is a significant relationship between a low standard of living that results from low income; and increased mortality rates. A low standard of living is more likely to result in malnutrition, which can make people more susceptible to disease and more likely to die from these diseases. A lower standard of living may lead to as a lack of hygiene and sanitation, increased exposure to and the spread of disease, and a lack of access to proper medical care and facilities. Poor health can in turn contribute to low and reduced incomes, which can create a loop known as the health-poverty trap. Indian economist and philosopher Amartya Sen has stated that mortality rates can serve as an indicator of economic success and failure.

Historically, mortality rates have been adversely affected by short term price increases. Studies have shown that mortality rates increase at a rate concurrent with increases in food prices. These effects have a greater impact on vulnerable, lower-income populations than they do on populations with a higher standard of living.

In more recent times, higher mortality rates have been less tied to socio-economic levels within a given society, but have differed more between low and high-income countries. It is now found that national income, which is directly tied to standard of living within a country, is the largest factor in mortality rates being higher in low-income countries.

Preventable Mortality

These rates are especially pronounced for children under 5 years old, particularly in lower-income, developing countries. These children have a much greater chance of dying of diseases that have become mostly preventable in higher-income parts of the world. More children die of malaria, respiratory infections, diarrhea, perinatal conditions, and measles in developing nations. Data shows that after the age of 5 these preventable causes level out between high and low-income countries.

19.2 LIFE TABLE

In actuarial science and demography, a life table (also called a mortality table or actuarial table) is a table which shows, for each age, the probability that a person of that age will die before their next birthday ("probability of death"). In other words, it represents the survivorship of people from a certain population. They can also be explained as a long-term mathematical way to measure a population's longevity.

Tables have been created by demographers including John Graunt, Reed and Merrell, Keyfitz, and Greville.

There are two types of life tables used in actuarial science. The period life table represents mortality rates during a specific time period for a certain population. A cohort life table, often referred to as a generation life table, is used to represent the overall mortality rates of a certain population's entire lifetime.

They must have had to be born during the same specific time interval. A cohort life table is more frequently used because it is able to make a prediction of any expected changes in the mortality rates of a population in the future.

This type of table also analyzes patterns in mortality rates that can be observed over time. Both of these types of life tables are created based on an actual population from the present, as well as an educated prediction of the experience of a population in the near future.

In order to find the true life expectancy average, 100 years would need to pass and by then finding that data would be of no use as healthcare is continually advancing.

Other life tables in historical demography may be based on historical records, although these often undercount infants and understate infant mortality, on comparison with other regions with better records, and on mathematical adjustments for varying mortality levels and life expectancies at birth.

From this starting point, a number of inferences can be derived.

- The probability of surviving any particular year of age.
- The remaining life expectancy for people at different ages.

Life tables are also used extensively in biology and epidemiology. An area that uses this tool is Social Security. It examines the mortality rates of all the people who have Social Security to decide which actions to take. The concept is also of importance in product life cycle management. All mortality tables are specific to environmental and life circumstances, and are used to probabilistically determine expected maximum age within those environmental conditions.

19.2.1 Background

There are two types of life tables:

- **Period or static** life tables show the current probability of death (for people of different ages, in the current year).
- **Cohort** life tables show the probability of death of people from a given cohort (especially birth year) over the course of their lifetime.

Static life tables sample individuals assuming a stationary population with overlapping generations. "Static life tables" and "cohort life tables" will be identical if population is in equilibrium and environment does not change. If a population were to have a constant number of people each year, it would mean that the probabilities of death from the life table were completely accurate. Also, an exact number of 100,000 people were born each year with no immigration or emigration involved. "Life table" primarily refers to *period* life tables, as cohort life tables can only be constructed using data up to the current point, and distant projections for future mortality.

Life tables can be constructed using projections of future mortality rates, but more often they are a snapshot of age-specific mortality rates in the recent past, and do not necessarily purport to be projections.

For these reasons, the older ages represented in a life table may have a greater chance of not being representative of what lives at these ages may experience in future, as it is predicated on current advances in medicine, public health, and safety standards that did not exist in the early years of this cohort.

A life table is created by mortality rates and census figures from a certain population, ideally under a closed demographic system. This means that immigration and emigration do not exist when analyzing a cohort.

A closed demographic system assumes that migration flows are random and not significant, and that immigrants from other populations have the same risk of death as an individual from the new population. Another benefit from mortality tables is that they can be used to make predictions on demographics or different populations.

However, there are also weaknesses of the information displayed on life tables. One being that they do not state the overall health of the population. There is more than one disease present in the world, and a person can have more than one disease at different stages simultaneously, introducing the term comorbidity. Therefore, life tables also do not show the direct correlation of mortality and morbidity.

The life table observes the mortality experience of a single generation, consisting of 100,000 births, at every age number they can live through.

Life tables are usually constructed separately for men and for women because of their substantially different mortality rates. Other characteristics can also be used to distinguish different risks, such as smoking status, occupation, and socioeconomic class.

Life tables can be extended to include other information in addition to mortality, for instance health information to calculate health expectancy. Health expectancies such as disability-adjusted life year and Healthy Life Years are the remaining number of years a person can expect to live in a specific health state, such as free of disability. Two types of life tables are used to divide the life expectancy into life spent in various states:

- Multi-state life tables (also known as increment-decrements life tables) are based on transition rates in and out of the different states and to death.
- Prevalence-based life tables (also known as the Sullivan method) are based on external information on the proportion in each state. Life tables can also be extended to show life expectancies in different labor force states or marital status states.

Life tables that relate to maternal deaths and infant moralities are important, as they help form family planning programs that work with particular populations. They also help compare a country's average life expectancy with other countries. Comparing life expectancy globally helps countries understand why one country's life expectancy is rising substantially by looking at each other's healthcare, and adopting ideas to their own systems.

19.2.2 Insurance Applications

In order to price insurance products, and ensure the solvency of insurance companies through adequate reserves, actuaries must develop projections of future insured events (such as death, sickness, and disability). To do this, actuaries develop mathematical models of the rates and timing of the events. They do this by studying the incidence of these events in the recent past, and sometimes developing expectations of how these past events will change over time (for example, whether the progressive reductions in mortality rates in the past will continue) and deriving expected rates of such events in the future, usually based on the age or other relevant characteristics of the population. An actuary's job is to form a comparison between people at risk of death and people who actually died to come up with a probability of death for a person at each age number, defined as qx in an equation. When analyzing a population, one of the main sources used to gather the required information is insurance by obtaining individual records that belong to a specific population. These are called mortality tables if they show death rates, and morbidity tables if they show various types of sickness or disability rates.

The availability of computers and the proliferation of data gathering about individuals has made possible calculations that are more voluminous and intensive than those used in the past (i.e. they crunch more numbers) and it is more common to attempt to provide different tables for different uses, and to factor in a range of non-traditional behaviors (e.g. gambling, debt load) into specialized calculations utilized by some institutions for evaluating risk.

This is particularly the case in non-life insurance (e.g. the pricing of motor insurance can allow for a large number of risk factors, which requires a correspondingly complex table of expected claim rates). However the expression "life table" normally refers to human survival rates and is not relevant to non-life insurance.

19.2.3 Ending a Mortality Table

In practice, it is useful to have an ultimate age associated with a mortality table. Once the ultimate age is reached, the mortality rate is assumed to be 1.000. This age may be the point at which life insurance benefits are paid to a survivor or annuity payments cease.

Four methods can be used to end mortality tables:

- **The Forced Method:** Select an ultimate age and set the mortality rate at that age equal to 1.000 without any changes to other mortality rates. This creates a discontinuity at the ultimate age compared to the penultimate and prior ages.
- **The Blended Method:** Select an ultimate age and blend the rates from some earlier age to dovetail smoothly into 1.000 at the ultimate age.
- **The Pattern Method:** Let the pattern of mortality continue until the rate approaches or hits 1.000 and set that as the ultimate age.
- **The Less-Than-One Method:** This is a variation on the Forced Method. The ultimate mortality rate is set equal to the expected mortality at a selected ultimate age, rather 1.000 as in the Forced Method. This rate will be less than 1.000.

19.2.4 Epidemiology

In epidemiology and public health, both standard life tables (used to calculate life expectancy), as well as the Sullivan and multi-state life tables (used to calculate health expectancy), are the most commonly mathematical used devices. The latter includes information on health in addition to mortality. By watching over the life expectancy of any year(s) being studied, epidemiologists can see if diseases are contributing to the overall increase in mortality rates.

Epidemiologists are able to help demographers understand the sudden decline of life expectancy by linking it to the health problems that are arising in certain populations.

CHAPTER 20
DEMOGRAPHY

Demography is the statistical study of human populations: their size, composition (e.g., ethnic group, age), and how they change through the interplay of fertility (births), mortality (deaths), and migration.

Demographic analysis examines and measures the dimensions and dynamics of populations; it can cover whole societies or groups defined by criteria such as education, nationality, religion, and ethnicity. Educational institutions usually treat demography as a field of sociology, though there are a number of independent demography departments.

These methods have primarily been developed to study human populations, but are extended to a variety of areas where researchers want to know how populations of social actors can change across time through processes of birth, death, and migration. In the context of human biological populations, demographic analysis uses administrative records to develop an independent estimate of the population. Demographic analysis estimates are often considered a reliable standard for judging the accuracy of the census information gathered at any time. In the labor force, demographic analysis is used to estimate sizes and flows of populations of workers; in population ecology the focus is on the birth, death, migration and immigration of individuals in a population of living organisms, alternatively, in social human sciences could involve movement of firms and institutional forms. Demographic analysis is used in a wide variety of contexts. For example, it is often used in business plans, to describe the population connected to the geographic location of the business. Demographic analysis is usually abbreviated as DA. For the 2010 U.S. Census, The U.S. Census Bureau has expanded its DA categories. Also as part of the 2010 U.S. Census, DA now also includes comparative analysis between independent housing estimates, and census address lists at different key time points.

Patient demographics form the core of the data for any medical institution, such as patient and emergency contact information and patient medical record data. They allow for the identification of a patient and their categorization into categories for the purpose of statistical analysis. Patient demographics include: date of birth, gender, date of death, postal code, ethnicity, blood type, emergency contact information, family doctor, insurance provider data, allergies, major diagnoses and major medical history.

Formal demography limits its object of study to the measurement of population processes, while the broader field of social demography or population studies also analyses the relationships between economic, social, institutional, cultural, and biological processes influencing a population.

20.1 HISTORY

Demographic thoughts traced back to antiquity, and were present in many civilisations and cultures, like Ancient Greece, Ancient Rome, China and India. Made up of the prefix *demo-* and the suffix *-graphy*, the term *demography* refers to the overall study of population.

In ancient Greece, this can be found in the writings of Herodotus, Thucydides, Hippocrates, Epicurus, Protagoras, Polus, Plato and Aristotle. In Rome, writers and philosophers like Cicero, Seneca, Pliny the Elder, Marcus Aurelius, Epictetus, Cato, and Columella also expressed important ideas on this ground.

In the Middle Ages, Christian thinkers devoted much time in refuting the Classical ideas on demography. Important contributors to the field were William of Conches, Bartholomew of Lucca, William of Auvergne, William of Pagula, and Muslim sociologists like Ibn Khaldun.

One of the earliest demographic studies in the modern period was *Natural and Political Observations Made upon the Bills of Mortality* (1662) by John Graunt, which contains a primitive form of life table. Among the study's findings were that one-third of the children in London died before their sixteenth birthday.

Mathematicians, such as Edmond Halley, developed the life table as the basis for life insurance mathematics. Richard Price was credited with the first textbook on life contingencies published in 1771, followed later by Augustus De Morgan, *On the Application of Probabilities to Life Contingencies* (1838).

In 1755, Benjamin Franklin published his essay *Observations Concerning the Increase of Mankind, Peopling of Countries, etc.*, projecting exponential growth in British colonies. His work influenced

Thomas Robert Malthus, who, writing at the end of the 18th century, feared that, if unchecked, population growth would tend to outstrip growth in food production, leading to ever-increasing famine and poverty. Malthus is seen as the intellectual father of ideas of overpopulation and the limits to growth. Later, more sophisticated and realistic models were presented by Benjamin Gompertz and Verhulst.

In 1855, a Belgian scholar Achille Guillard defined demography as the natural and social history of human species or the mathematical knowledge of populations, of their general changes, and of their physical, civil, intellectual, and moral condition. The period 1860–1910 can be characterized as a period of transition where in demography emerged from statistics as a separate field of interest. This period included a panoply of international 'great demographers' like Adolphe Quetelet (1796–1874), William Farr (1807–1883), Louis-Adolphe Bertillon (1821–1883) and his son Jacques (1851–1922), Joseph Körösi (1844–1906), Anders Nicolas Kaier (1838–1919), Richard Böckh (1824–1907), Émile Durkheim (1858–1917), Wilhelm Lexis (1837–1914), and Luigi Bodio (1840–1920) contributed to the development of demography and to the toolkit of methods and techniques of demographic analysis.

20.2 METHODS

Demography is the statistical and mathematical study of the size, composition, and spatial distribution of human populations and how these features change over time. Data are obtained from a census of the population and from registries: records of events like birth, deaths, migrations, marriages, divorces, diseases, and employment. To do this, there needs to be an understanding of how they are calculated and the questions they answer which are included in these four concepts: population change, standardization of population numbers, the demographic bookkeeping equation, and population composition. There are two types of data collection—direct and indirect—with several methods of each type.

20.2.1 Direct Methods

Direct data comes from vital statistics registries that track all births and deaths as well as certain changes in legal status such as marriage, divorce, and migration (registration of place of residence). In developed countries with good registration systems (such as the United States and much of Europe), registry statistics are the best method for estimating the number of births and deaths.

A census is the other common direct method of collecting demographic data. A census is usually conducted by a national government and attempts to enumerate every person in a country. In contrast to vital statistics data, which are typically collected continuously and summarized on an annual basis, censuses typically occur only every 10 years or so, and thus are not usually the best source of data on births and deaths. Analyses are conducted after a census to estimate how much over or undercounting took place. These compare the sex ratios from the census data to those estimated from natural values and mortality data.

Censuses do more than just count people. They typically collect information about families or households in addition to individual characteristics such as age, sex, marital status, literacy/education, employment status, and occupation, and geographical location. They may also collect data on migration (or place of birth or of previous residence), language, religion, nationality (or ethnicity or race), and citizenship. In countries in which the vital registration system may be incomplete, the censuses are also used as a direct source of information about fertility and mortality; for example, the censuses of the People's Republic of China gather information on births and deaths that occurred in the 18 months immediately preceding the census.

20.2.2 Indirect Methods

Indirect methods of collecting data are required in countries and periods where full data are not available, such as is the case in much of the developing world, and most of historical demography. One of these techniques in contemporary demography is the sister method, where survey researchers ask women how many of their sisters have died or had children and at what age. With these surveys, researchers can then indirectly estimate birth or death rates for the entire population. Other indirect methods in contemporary demography include asking people about siblings, parents, and children. Other indirect methods are necessary in historical demography.

There are a variety of demographic methods for modelling population processes. They include models of mortality (including the life table, Gompertz models, hazards models, Cox proportional hazards models, multiple decrement life tables, Brass relational logits), fertility (Hermes model, Coale-Trussell models, parity progression ratios), marriage (Singulate Mean at Marriage, Page model), disability (Sullivan's method, multistate life tables), population projections (Lee-Carter model, the Leslie Matrix), and population momentum (Keyfitz).

The United Kingdom has a series of four national birth cohort studies, the first three spaced apart by 12 years: the 1946 National Survey of Health and Development, the 1958 National Child Development Study, the 1970 British Cohort Study, and the Millennium Cohort Study, begun much more recently in 2000. These have followed the lives of samples of people (typically beginning with around 17,000 in each study) for many years, and are still continuing. As the samples have been drawn in a nationally representative way, inferences can be drawn from these studies about the differences between four distinct generations of British people in terms of their health, education, attitudes, childbearing and employment patterns.

Indirect standardization is used when a population is small enough that the number of events (births, deaths, etc.) are also small. In this case, methods must be used to produce a standardized mortality rate (SMR) or standardized incidence rate (SIR).

20.3 POPULATION CHANGE

Population change is analyzed by measuring the change between one population size to another. Global population continues to rise, which makes population change an essential component to demographics. This is calculated by taking one population size minus the population size in an earlier census. The best way of measuring population change is using the intercensal percentage change. The intercensal percentage change is the absolute change in population between the censuses divided by the population size in the earlier census. Next, multiply this a hundredfold to receive a percentage. When this statistic is achieved, the population growth between two or more nations that differ in size, can be accurately measured and examined.

20.4 STANDARDIZATION OF POPULATION NUMBERS

For there to be a significant comparison, numbers must be altered for the size of the population that is under study. For example, the fertility rate is calculated as the ratio of the number of births to women of childbearing age to the total number of women in this age range. If these adjustments were not made, we would not know if a nation with a higher rate of births or deaths has a population with more women of childbearing age or more births per eligible woman. Within the category of standardization, there are two major approaches: direct standardization and indirect standardization.

20.5 COMMON RATES AND RATIOS

- The crude birth rate, the annual number of live births per 1,000 people.
- The general fertility rate, the annual number of live births per 1,000 women of childbearing age (often taken to be from 15 to 49 years old, but sometimes from 15 to 44).
- The age-specific fertility rates, the annual number of live births per 1,000 women in particular age groups (usually age 15–19, 20–24 etc.)
- The crude death rate, the annual number of deaths per 1,000 people.
- The infant mortality rate, the annual number of deaths of children less than 1 year old per 1,000 live births.
- The expectation of life (or life expectancy), the number of years that an individual at a given age could expect to live at present mortality levels.
- The total fertility rate, the number of live births per woman completing her reproductive life, if her childbearing at each age reflected current age-specific fertility rates.
- The replacement level fertility, the average number of children women must have in order to replace the population for the next generation. For example, the replacement level fertility in the US is 2.11.
- The gross reproduction rate, the number of daughters who would be born to a woman completing her reproductive life at current age-specific fertility rates.

- The net reproduction ratio is the expected number of daughters, per newborn prospective mother, who may or may not survive to and through the ages of childbearing.
- A stable population, one that has had constant crude birth and death rates for such a long period of time that the percentage of people in every age class remains constant, or equivalently, the population pyramid has an unchanging structure.
- A stationary population, one that is both stable and unchanging in size (the difference between crude birth rate and crude death rate is zero).
- Measures of centralisation are concerned with the extent to which an area's population is concentrated in its urban centres.

A stable population does not necessarily remain fixed in size. It can be expanding or shrinking. The crude death rate as defined above and applied to a whole population can give a misleading impression. For example, the number of deaths per 1,000 people can be higher in developed nations than in less-developed countries, despite standards of health being better in developed countries. This is because developed countries have proportionally more older people, who are more likely to die in a given year, so that the overall mortality rate can be higher even if the mortality rate at any given age is lower. A more complete picture of mortality is given by a life table, which summarizes mortality separately at each age. A life table is necessary to give a good estimate of life expectancy.

20.6 BASIC EQUATION REGARDING DEVELOPMENT OF A POPULATION

Suppose that a country (or other entity) contains $Population_t$ persons at time t. What is the size of the population at time $t + 1$?

$$\text{Population}_{t+1} = \text{Population}_t + \text{Natural Increase}_t + \text{Net Migration}_t$$

Natural increase from time t to $t + 1$:

$$\text{Natural Increase}_t = \text{Births}_t - \text{Deaths}_t$$

Net migration from time t to $t + 1$:

$$\text{Net Migration}_t = \text{Immigration}_t - \text{Emigration}_t$$

These basic equations can also be applied to subpopulations. For example, the population size of ethnic groups or nationalities within a given society or country is subject to the same sources of change. When dealing with ethnic groups, however, "net migration" might have to be subdivided into physical migration and ethnic reidentification (assimilation). Individuals who change their ethnic self-labels or whose ethnic classification in government statistics changes over time may be thought of as migrating or moving from one population subcategory to another.

More generally, while the basic demographic equation holds true by definition, in practice the recording and counting of events (births, deaths, immigration, emigration) and the enumeration of the total population size are subject to error. So allowance needs to be made for error in the underlying statistics when any accounting of population size or change is made.

The figure in this section shows the latest (2004) UN (United Nations) WHO projections of world population out to the year 2150 (red = high, orange = medium, green = low). The UN "medium" projection shows world population reaching an approximate equilibrium at 9 billion by 2075. Working independently, demographers at the International Institute for Applied Systems Analysis in Austria expect world population to peak at 9 billion by 2070. Throughout the 21st century, the average age of the population is likely to continue to rise.

20.6.1 Science of Population

Populations can change through three processes: fertility, mortality, and migration. Fertility involves the number of children that women have and is to be contrasted with fecundity (a woman's childbearing potential). Mortality is the study of the causes, consequences, and measurement of processes affecting death to members of the population. Demographers most commonly study mortality using the life table, a statistical device that provides information about the mortality conditions (most notably the life expectancy) in the population.

Migration refers to the movement of persons from a locality of origin to a destination place across some predefined, political boundary. Migration researchers do not designate movements 'migrations' unless they are somewhat permanent. Thus, demographers do not consider tourists and travellers to be migrating. While demographers who study migration typically do so through census data on place of residence, indirect sources of data including tax forms and labour force surveys are also important.

Demography is today widely taught in many universities across the world, attracting students with initial training in social sciences, statistics or health studies. Being at the crossroads of several disciplines such as sociology, economics, epidemiology, geography, anthropology and history, demography offers tools to approach a large range of population issues by combining a more technical quantitative approach that represents the core of the discipline with many other methods borrowed from social or other sciences. Demographic research is conducted in universities, in research institutes, as well as in statistical departments and in several international agencies. Population institutions are part of the CICRED (International Committee for Coordination of Demographic Research) network while most individual scientists engaged in demographic research are members of the International

Union for the Scientific Study of Population, or a national association such as the Population Association of America in the United States, or affiliates of the Federation of Canadian Demographers in Canada.

20.7 POPULATION COMPOSITION

Population composition is the description of population defined by characteristics such as age, race, sex or marital status. These descriptions can be necessary for understanding the social dynamics from historical and comparative research. This data is often compared using a population pyramid.

Population composition is also a very important part of historical research. Information ranging back hundreds of years is not always worthwhile, because the numbers of people for which data are available may not provide the information that is important (such as population size). Lack of information on the original data-collection procedures may prevent accurate evaluation of data quality.

20.8 DEMOGRAPHIC ANALYSIS IN INSTITUTIONS AND ORGANIZATIONS

20.8.1 Labor Market

The demographic analysis of labor markets can be used to show slow population growth, population aging, and the increased importance of immigration. The U.S. Census Bureau projects that in the next 100 years, the United States will face some dramatic demographic changes. The population is expected to grow more slowly and age more rapidly than ever before and the nation will become a nation of immigrants. This influx is projected to rise over the next century as new immigrants and their children will account for over half the U.S. population. These demographic shifts could ignite major adjustments in the economy, more specifically, in labor markets.

20.8.2 Turnover and in Internal Labor Markets

People decide to exit organizations for many reasons, such as, better jobs, dissatisfaction, and concerns within the family. The causes of turnover can be split into two separate factors, one linked with the culture of the organization, and the other relating to all other factors. People who do not fully accept a culture might leave voluntarily. Or, some individuals might leave because they fail to fit in and fail to change within a particular organization.

20.8.3 Population Ecology of Organizations

A basic definition of population ecology is a study of the distribution and abundance of organisms. As it relates to organizations and demography, organizations go through various liabilities to their continued survival. Hospitals, like all other large and complex organizations are impacted in the environment they work. For example, a study was done on the closure of acute care hospitals in Florida between a particular time. The study examined effect size, age, and niche density of these particular hospitals. A population theory says that organizational outcomes are mostly determined by environmental factors. Among several factors of the theory, there are four that apply to the hospital closure example: size, age, density of niches in which organizations operate, and density of niches in which organizations are established.

Business Organizations

Problems in which demographers may be called upon to assist business organizations are when determining the best prospective location in an area of a branch store or service outlet, predicting the demand for a new product, and to analyze certain dynamics of a company's workforce. Choosing a new location for a branch of a bank, choosing the area in which to start a new supermarket, consulting a bank loan officer that a particular location would be a beneficial site to start a car wash, and determining what shopping area would be best to buy and be redeveloped in metropolis area are types of problems in which demographers can be called upon. Standardization is a useful demographic technique used in the analysis of a business. It can be used as an interpretive and analytic tool for the comparison of different markets.

Nonprofit Organizations

These organizations have interests about the number and characteristics of their clients so they can maximize the sale of their products, their outlook on their influence, or the ends of their power, services, and beneficial works.

CHAPTER 21

INTRODUCTION TO RESEARCH METHODS

Research methods are the strategies, techniques, and processes used to collect and analyze information in order to answer questions or test hypotheses. These methods provide the tools needed to gather reliable and valid data, interpret results, and contribute to knowledge in various fields such as science, social science, medicine, business, and education. Selecting the right research methods is essential for the success of any research project, as they influence the accuracy and credibility of the findings.

Research methods can be broadly divided into two categories: qualitative and quantitative. Qualitative methods focus on understanding phenomena through in-depth observation, interviews, and analysis of non-numerical data, while quantitative methods emphasize numerical measurement and statistical analysis. Both types of methods have their unique strengths and are often used in combination to provide a more comprehensive understanding of the research problem.

21.1 TYPES OF RESEARCH METHODS

21.1.1 Qualitative Research Methods

Qualitative research methods are designed to explore complex phenomena by focusing on the meanings, experiences, and perspectives of individuals. These methods are often used in social sciences, humanities, and health research to provide rich, descriptive insights into behaviors, emotions, and social interactions. Unlike quantitative methods, which rely on numerical data, qualitative research emphasizes depth over breadth.

Some common qualitative research methods include:

- **Interviews:** One-on-one interviews involve asking participants open-ended questions to explore their thoughts, experiences, and motivations. Interviews can be structured, semi-structured, or unstructured, depending on the level of control the researcher wishes to maintain over the conversation.
- **Focus Groups:** A focus group consists of a small group of people who discuss a particular topic or issue in a moderated setting. This method

allows researchers to gather multiple perspectives and observe group dynamics.

- **Observation:** Researchers may use observation to study participants in their natural environment. This method is useful for understanding behaviors and interactions that may not be captured through interviews or surveys.
- **Case Studies:** A case study involves an in-depth examination of a single individual, group, or event. This method provides detailed insights into specific cases and can generate hypotheses for further research.

Qualitative research methods are typically flexible and adaptable, allowing researchers to explore issues in real-time and in real-world settings. However, they are often criticized for their lack of generalizability, as findings from small, non-representative samples may not apply to larger populations.

21.1.2 Quantitative Research Methods

Quantitative research methods focus on numerical measurement and statistical analysis. These methods are used to test hypotheses, identify patterns, and make predictions based on data. Quantitative research is often seen as more objective than qualitative research because it relies on measurable, reproducible data.

Some common quantitative research methods include:

- **Surveys:** Surveys are one of the most widely used quantitative research methods. Researchers use questionnaires to collect data from a large sample of participants. Surveys can be administered in person, by mail, online, or via telephone, and they often involve closed-ended questions that can be easily quantified.
- **Experiments:** In experimental research, researchers manipulate one or more variables to determine their effect on a particular outcome. Experiments typically involve a control group and an experimental group to isolate the effect of the independent variable. Randomized controlled trials (RCTs) are a common type of experiment used in fields such as medicine and psychology.
- **Secondary Data Analysis:** Researchers can use existing datasets (e.g., census data, health records, or economic data) to conduct secondary data analysis. This method is often more cost-effective than collecting new data and allows researchers to analyze large, representative samples.
- **Content Analysis:** Content analysis is a quantitative method used to analyze textual, visual, or media content. Researchers code and categorize information to identify patterns, trends, or correlations.

Quantitative research methods offer several advantages, including the ability to generalize findings to larger populations and the use of rigorous statistical techniques to ensure the reliability and validity of the results. However, these methods may overlook the nuances of human behavior and experience, which are often better captured through qualitative methods.

21.1.3 Mixed Methods Research

Mixed methods research combines both qualitative and quantitative approaches to provide a more comprehensive understanding of a research problem. This approach allows researchers to capitalize on the strengths of each method while minimizing their respective weaknesses. For example, a researcher may use qualitative methods to explore a phenomenon in-depth and then use quantitative methods to test the generalizability of the findings.

Mixed methods research can be conducted sequentially or concurrently. In a **sequential approach**, one method is used first, followed by the other (e.g., conducting interviews to develop a survey). In a **concurrent approach**, both methods are used simultaneously, and the results are integrated during the analysis phase. Mixed methods research is becoming increasingly popular because it provides a more nuanced understanding of complex research problems. However, it requires expertise in both qualitative and quantitative techniques, and the integration of data from both approaches can be challenging.

21.2 KEY STEPS IN THE RESEARCH PROCESS

21.2.1 Defining the Research Question

The first step in any research project is to define the research question or problem. A well-defined research question provides focus and direction for the study. It should be specific, measurable, and researchable within the constraints of time, resources, and ethical considerations. In qualitative research, the research question may evolve as the study progresses, while in quantitative research, it is typically defined upfront and remains fixed.

21.2.2 Literature Review

Before conducting original research, it is essential to review existing literature on the topic. A literature review helps researchers understand what has already been studied, identify gaps in knowledge, and refine the research question. It also provides a theoretical framework for the study and helps researchers avoid duplicating previous work.

21.2.3 Designing the Study

Once the research question is defined and the literature has been reviewed, researchers must design the study. This involves selecting the appropriate research methods, determining the sample size, and deciding how data will be collected and analyzed. In quantitative research, this step often involves developing a detailed research protocol, while in qualitative research, the design may be more flexible and adaptive.

21.2.4 Data Collection

Data collection is the process of gathering information to answer the research question.

Depending on the research method, data may be collected through surveys, interviews, observations, or experiments. Researchers must ensure that data collection procedures are ethical and that participants provide informed consent. Additionally, researchers must minimize bias in data collection to ensure that the findings are valid and reliable.

21.2.5 Data Analysis

After data collection, researchers must analyze the data to draw meaningful conclusions.

In qualitative research, data analysis often involves coding and categorizing themes, while in quantitative research, statistical techniques are used to test hypotheses and identify relationships between variables. The choice of data analysis methods depends on the type of data collected and the research question.

21.2.6 Interpreting and Reporting Results

The final step in the research process is to interpret the results and report the findings. Researchers must carefully consider the implications of their findings and discuss how they relate to the research question and existing literature. In quantitative research, this may involve discussing the statistical significance of the results, while in qualitative research, it may involve exploring the broader meaning of the findings.

Researchers typically report their results in the form of research papers, presentations, or reports. Clear and transparent reporting is essential for ensuring that others can evaluate and replicate the study if necessary.

21.3 ETHICAL CONSIDERATIONS IN RESEARCH

Ethics play a crucial role in research, especially when it involves human participants. Researchers must obtain informed consent from participants, ensure confidentiality, and avoid causing harm. In addition, researchers must be transparent about their methods and report their findings honestly. Ethical research is essential for maintaining public trust and ensuring the integrity of the research process. In fields such as medicine and psychology, research involving human subjects is often reviewed by institutional review boards (IRBs) or ethics committees to ensure that ethical standards are met.

21.4 CONCLUSION

Research methods are essential tools for answering important questions and advancing knowledge in a wide range of disciplines. Whether qualitative, quantitative, or mixed methods are used, careful planning and execution are crucial to the success of any research project. By selecting appropriate methods, collecting valid data, and analyzing results accurately, researchers can make meaningful contributions to their fields and provide evidence-based solutions to real-world problems. Furthermore, adherence to ethical standards ensures that research is conducted responsibly and with respect for the rights and well-being of participants.

CHAPTER 22
INTERVENTIONAL STUDIES

Interventional studies are a critical component of clinical research, designed to evaluate the effects of specific interventions on health outcomes. These studies are essential for advancing medical knowledge, developing new treatments, and improving patient care. By actively manipulating variables, interventional studies provide a robust framework for understanding the causal relationships between interventions and their effects on participants.

22.1 TYPES OF INTERVENTIONAL STUDIES

Interventional studies can be categorized into several types, each with distinct features and applications. The most common types include randomized controlled trials (RCTs), non-randomized controlled trials, and cohort studies.

22.1.1 Randomized Controlled Trials (RCTs)

RCTs are considered the gold standard in interventional research due to their ability to minimize bias and establish causal relationships. In an RCT, participants are randomly assigned to either the intervention group, which receives the treatment, or the control group, which may receive a placebo or standard care. Randomization helps ensure that the groups are comparable and that any observed differences in outcomes can be attributed to the intervention itself.

The RCT design includes several key components, such as blinding, sample size determination, and outcome measurement. Blinding, or masking, is often employed to prevent bias in outcome assessment.

In single-blind studies, participants are unaware of their group assignment, while in double-blind studies, both participants and researchers are blinded. Proper sample size calculation is critical to ensure sufficient power to detect meaningful differences between groups.

22.1.2 Non-Randomized Controlled Trials

In some cases, randomization may not be feasible or ethical. Non-randomized controlled trials, also known as quasi-experimental studies, do not use random assignment to allocate participants to intervention and control groups. Instead, these studies rely on other methods, such as matching or statistical adjustments, to control for confounding variables.

While non-randomized trials can provide valuable insights, they are more susceptible to bias compared to RCTs, making it more challenging to draw definitive conclusions about causality.

22.1.3 Cohort Studies

Cohort studies, though not strictly interventional, can provide important information about the effects of interventions in real-world settings. In these studies, groups of individuals (cohorts) who have been exposed to an intervention are compared to those who have not. While cohort studies do not involve randomization, they can still offer valuable insights into the effectiveness and safety of interventions over time. Cohort studies are particularly useful for evaluating interventions in large populations or when RCTs are not feasible due to ethical or logistical reasons.

22.2 DESIGN AND METHODOLOGY OF INTERVENTIONAL STUDIES

The design of interventional studies is crucial for ensuring their validity and reliability. Several key elements must be considered when designing these studies:

22.2.1 Defining the Research Question

The first step in designing an interventional study is to clearly define the research question. This involves specifying the intervention, population, and outcomes of interest. A well-defined research question guides the study design and ensures that the research objectives are aligned with the needs of the target population.

22.2.2 Selecting the Intervention

The choice of intervention is critical to the success of the study. Interventions can include pharmacological treatments, behavioral interventions, surgical procedures, or educational programs. The intervention must be clearly defined, including its dosage, duration, and mode of delivery. Additionally, the intervention should be feasible and acceptable to participants.

22.2.3 Sample Size Calculation

Determining the appropriate sample size is essential for ensuring that the study has enough power to detect meaningful differences between groups. Sample size calculations consider factors such as the expected effect size, variability in the outcome measures, and the significance level. A larger sample size increases the reliability of the results but also increases the resources required for the study.

22.2.4 Randomization and Blinding

In RCTs, randomization is a critical element that helps control for bias. Researchers can use various methods for randomization, including simple random sampling, stratified random sampling, and block randomization. Blinding, as mentioned earlier, helps reduce bias in outcome assessment. Both randomization and blinding are essential for maintaining the integrity of the study.

22.2.5 Data Collection and Outcome Measurement

Accurate data collection is vital for assessing the effectiveness of the intervention. Researchers must select appropriate outcome measures that are valid, reliable, and relevant to the research question. Outcomes can be primary (the main focus of the study) or secondary (additional outcomes of interest). Data collection methods may include surveys, clinical assessments, laboratory tests, and patient-reported outcomes.

22.3 ETHICAL CONSIDERATIONS IN INTERVENTIONAL STUDIES

Ethical considerations are paramount in the conduct of interventional studies, particularly when human participants are involved. Researchers must adhere to ethical guidelines to ensure the rights and welfare of participants. Key ethical principles include:

22.3.1 Informed Consent

Participants must provide informed consent before enrolling in an interventional study. This process involves informing them about the study's purpose, procedures, risks, benefits, and their right to withdraw at any time. Informed consent ensures that participants are fully aware of what they are agreeing to and can make an autonomous decision.

22.3.2 Minimizing Harm

Researchers must take measures to minimize harm to participants. This includes conducting a thorough risk-benefit analysis to ensure that the potential benefits of the intervention outweigh the risks. Ongoing monitoring of participants during the study is essential to detect and address any adverse effects.

22.3.3 Ethical Review

Interventional studies are typically subject to review by an institutional review board (IRB) or ethics committee. These bodies evaluate the study's ethical considerations, including the research design, informed consent process, and data management procedures. Approval from an IRB is essential before commencing any study involving human subjects.

22.4 CHALLENGES IN CONDUCTING INTERVENTIONAL STUDIES

Despite their importance, interventional studies face several challenges that can impact their design and execution:

22.4.1 Recruitment and Retention

Recruiting and retaining participants can be challenging, particularly in studies with strict eligibility criteria. Researchers must develop effective recruitment strategies and provide incentives to encourage participation. Additionally, strategies must be implemented to minimize dropout rates and ensure that participants remain engaged throughout the study.

22.4.2 Generalizability

While RCTs are designed to control for bias, the findings may not always be generalizable to the broader population. Participants in RCTs may differ from the general population in terms of demographics, health status, or comorbidities. Researchers must consider the external validity of their findings and how they may apply to different populations.

22.4.3 Compliance and Adherence

Ensuring that participants adhere to the intervention protocol is crucial for the integrity of the study. Non-compliance can introduce bias and affect the validity of the results. Researchers can use strategies such as reminders, education, and support to enhance adherence to the intervention.

22.5 SIGNIFICANCE OF INTERVENTIONAL STUDIES IN EVIDENCE-BASED MEDICINE

Interventional studies play a vital role in evidence-based medicine by providing high-quality data that informs clinical practice. The findings from interventional studies contribute to the development of clinical guidelines, treatment protocols, and public health policies. They enable healthcare professionals to make informed decisions based on the best available evidence, ultimately improving patient outcomes and quality of care. Furthermore, interventional studies are essential for identifying effective treatments and interventions for various health conditions. By rigorously evaluating the benefits and risks of interventions, researchers can ensure that patients receive safe and effective care.

22.6 CONCLUSION

Interventional studies are a cornerstone of clinical research, providing valuable insights into the effects of specific interventions on health outcomes. Through careful design, ethical considerations, and rigorous methodology, these studies contribute significantly to evidence-based medicine and the advancement of healthcare. Despite the challenges they face, interventional studies remain essential for developing new treatments, improving patient care, and addressing public health issues. As research continues to evolve, the importance of interventional studies in shaping the future of medicine cannot be overstated.

CHAPTER 23
SAMPLE SIZE DETERMINATION

Sample size determination or estimation is the act of choosing the number of observations or replicates to include in a statistical sample. The sample size is an important feature of any empirical study in which the goal is to make inferences about a population from a sample. In practice, the sample size used in a study is usually determined based on the cost, time, or convenience of collecting the data, and the need for it to offer sufficient statistical power.

In complex studies, different sample sizes may be allocated, such as in stratified surveys or experimental designs with multiple treatment groups. In a census, data is sought for an entire population, hence the intended sample size is equal to the population.

In experimental design, where a study may be divided into different treatment groups, there may be different sample sizes for each group.

Sample sizes may be chosen in several ways:

- Using experience—small samples, though sometimes unavoidable, can result in wide confidence intervals and risk of errors in statistical hypothesis testing.
- Using a target variance for an estimate to be derived from the sample eventually obtained, i.e., if a high precision is required (narrow confidence interval) this translates to a low target variance of the estimator.
- The use of a power target, i.e. the power of statistical test to be applied once the sample is collected.
- Using a confidence level, i.e. the larger the required confidence level, the larger the sample size (given a constant precision requirement).

23.1 INTRODUCTION

Sample size determination is a crucial aspect of research methodology that plays a significant role in ensuring the reliability and validity of study findings. In order to influence the accuracy of estimates, the power of statistical tests, and the general robustness of the research findings, it entails carefully choosing the number of participants or data points to be included in a study.

For example, if we are conducting a survey to determine the average satisfaction level of customers regarding a new product. To determine an appropriate sample size, we need to consider factors such as the desired level of confidence, margin of error, and variability in the responses. We might decide that we want a 95% confidence level, meaning we are 95% confident that the true average satisfaction level falls within the calculated range. We also decide on a margin of error, of ±3%, which indicates the acceptable range of difference between our sample estimate and the true population parameter. Additionally, we may have some idea of the expected variability in satisfaction levels based on previous data or assumptions.

23.2 IMPORTANCE

Larger sample sizes generally lead to increased precision when estimating unknown parameters. For instance, to accurately determine the prevalence of pathogen infection in a specific species of fish, it is preferable to examine a sample of 200 fish rather than 100 fish. Several fundamental facts of mathematical statistics describe this phenomenon, including the law of large numbers and the central limit theorem.

In some situations, the increase in precision for larger sample sizes is minimal, or even non-existent. This can result from the presence of systematic errors or strong dependence in the data, or if the data follows a heavy-tailed distribution, or because the data is strongly dependent or biased.

Sample sizes may be evaluated by the quality of the resulting estimates, as follows. It is usually determined on the basis of the cost, time or convenience of data collection and the need for sufficient statistical power. For example, if a proportion is being estimated, one may wish to have the 95% confidence interval be less than 0.06 units wide. Alternatively, sample size may be assessed based on the power of a hypothesis test. For example, if we are comparing the support for a certain political candidate among women with the support for that candidate among men, we may wish to have 80% power to detect a difference in the support levels of 0.04 units.

23.3 ESTIMATION

23.3.1 Estimation of a Proportion

A relatively simple situation is estimation of a proportion. It is a fundamental aspect of statistical analysis, particularly when gauging the prevalence of a specific characteristic within a population. For example, we may wish to estimate the proportion of residents in a community who are at least 65 years old.

The estimator of a proportion is $\hat{p} = X / n$, where X is the number of 'positive' instances (e.g., the number of people out of the n sampled people who are at least 65 years old). When the observations are independent, this estimator has a (scaled) binomial distribution (and is also the sample mean of data from a Bernoulli distribution). The maximum variance of this distribution is 0.25, which occurs when the true parameter is $p = 0.5$. In practical applications, where the true parameter p is unknown, the maximum variance is often employed for sample size assessments. If a reasonable estimate for p is known the quantity $p(1-p)$ may be used in place of 0.25.

As the sample size n grows sufficiently large, the distribution of $\hat{p}$ will be closely approximated by a normal distribution. Using this and the Wald method for the binomial distribution, yields a confidence interval, with Z representing the standard Z-score for the desired confidence level (e.g., 1.96 for a 95% confidence interval), in the form:

$$\left(\hat{p} - Z\sqrt{\frac{0.25}{n}},\quad \hat{p} + Z\sqrt{\frac{0.25}{n}}\right)$$

To determine an appropriate sample size n for estimating proportions, the equation below can be solved, where W represents the desired width of the confidence interval. The resulting sample size formula, is often applied with a conservative estimate of p (e.g., 0.5):

$$Z\sqrt{\frac{0.25}{n}} = W/2$$

For n, yielding the sample size

$$n = \frac{Z^2}{W^2}$$

in the case of using .5 as the most conservative estimate of the proportion. *(Note: W/2 = margin of error.)*

In the figure below one can observe how sample sizes for binomial proportions change given different confidence levels and margins of error.

Otherwise, the formula would be

$$Z\sqrt{\frac{p(1-p)}{n}} = W/2$$

which yields

$$n = \frac{4Z^2 p(1-p)}{W^2}$$

For example, in estimating the proportion of the U.S. population supporting a presidential candidate with a 95% confidence interval width of 2 percentage points (0.02), a sample size of $(1.96)^2/(0.02^2) = 9604$ is required with the margin of error in this case is 1 percentage point. It is reasonable to use the 0.5 estimate for p in this case because the presidential races are often close to 50/50, and it is also prudent to use a conservative estimate. The margin of error in this case is 1 percentage point (half of 0.02).

In practice, the formula :

$$\left(\hat{p}-1.96\sqrt{\frac{0.25}{n}},\quad \hat{p}+1.96\sqrt{\frac{0.25}{n}}\right)$$

is commonly used to form a 95% confidence interval for the true proportion. The equation

$$4\sqrt{\frac{0.25}{n}}=W/2$$

can be solved for n, providing a minimum sample size needed to meet the desired margin of error. The foregoing is commonly simplified: $n = 4/W^2 = 1/B^2$ where B is the error bound on the estimate, i.e., the estimate is usually given as *within* $\pm B$. For $B = 10\%$ one requires $n = 100$, for $B = 5\%$ one needs $n = 400$, for $B = 3\%$ the requirement approximates to $n = 1000$, while for $B = 1\%$ a sample size of $n = 10000$ is required. These numbers are quoted often in news reports of opinion polls and other sample surveys.

However, the results reported may not be the exact value as numbers are preferably rounded up. Knowing that the value of the n is the minimum number of sample points needed to acquire the desired result, the number of respondents then must lie on or above the minimum.

23.3.2 Estimation of a Mean

Simply speaking, if we are trying to estimate the average time it takes for people to commute to work in a city. Instead of surveying the entire population, you can take a random sample of 100 individuals, record their commute times, and then calculate the mean (average) commute time for that sample.

For example, person 1 takes 25 minutes, person 2 takes 30 minutes, ..., person 100 takes 20 minutes. Add up all the commute times and divide by the number of people in the sample (100 in this case).

The result would be your estimate of the mean commute time for the entire population. This method is practical when it's not feasible to measure everyone in the population, and it provides a reasonable approximation based on a representative sample.

In a precisely mathematical way, when estimating the population mean using an independent and identically distributed (iid) sample of size n, where each data value has variance σ^2, the standard error of the sample mean is:

$$\frac{\sigma}{\sqrt{n}}.$$

This expression describes quantitatively how the estimate becomes more precise as the sample size increases. Using the central limit theorem to justify approximating the sample mean with a normal distribution yields a confidence interval of the form

$$\left(\bar{x}-\frac{Z\sigma}{\sqrt{n}},\quad \bar{x}+\frac{Z\sigma}{\sqrt{n}}\right)$$

Where Z is a standard Z-score for the desired level of confidence (1.96 for a 95% confidence interval).

To determine the sample size *n* required for a confidence interval of width W, with

$$\frac{Z\sigma}{\sqrt{n}}=W/2$$

W/2 as the margin of error on each side of the sample mean, the equation can be solved. This yields the sample size formula, for *n*:

$$n=\frac{4Z^2\sigma^2}{W^2}$$

For instance, if estimating the effect of a drug on blood pressure with a 95% confidence interval that is six units wide, and the known standard deviation of

$$\frac{4\times1.96^2\times15^2}{6^2}=96.04$$

blood pressure in the population is 15, the required sample size would be , which would be rounded up to 97, since sample sizes must be integers and must meet or exceed the calculated *minimum* value. Understanding these calculations is essential for researchers designing studies to accurately estimate population means within a desired level of confidence.

23.4 REQUIRED SAMPLE SIZES FOR HYPOTHESIS TESTS

One of the prevalent challenges faced by statisticians revolves around the task of calculating the sample size needed to attain a specified statistical power for a test, all while maintaining a pre-determined Type I error rate α, which signifies the level of significance in hypothesis testing. It yields a certain power for a test, given a predetermined. As follows, this can be estimated by pre-determined tables for certain values, by Mead's resource equation, or, more generally, by the cumulative distribution function:

23.4.1 Tables

The table can be used in a two-sample t-test to estimate the sample sizes of an experimental group and a control group that are of equal size, that is, the total number of individuals in the trial is twice that of the number given, and the desired significance level is 0.05.

Power	0.2	0.5	0.8
0.25	84	14	6
0.50	193	32	13
0.60	246	40	16
0.70	310	50	20
0.80	393	64	26
0.90	526	85	34
0.95	651	105	42
0.99	920	148	58

The parameters used are:

- The desired statistical power of the trial, shown in column to the left.
- Cohen's d (= effect size), which is the expected difference between the means of the target values between the experimental group and the control group, divided by the expected standard deviation.

23.4.2 Mead's Resource Equation

Mead's resource equation is often used for estimating sample sizes of laboratory animals, as well as in many other laboratory experiments. It may not be as accurate as using other methods in estimating sample size, but gives a hint of what is the appropriate sample size where parameters such as expected standard deviations or expected differences in values between groups are unknown or very hard to estimate.

All the parameters in the equation are in fact the degrees of freedom of the number of their concepts, and hence, their numbers are subtracted by 1 before insertion into the equation.

The equation is:

$$E = N - B - T,$$

where:

- N is the total number of individuals or units in the study (minus 1)
- B is the *blocking component*, representing environmental effects allowed for in the design (minus 1)
- T is the *treatment component*, corresponding to the number of treatment groups (including control group) being used, or the number of questions being asked (minus 1)

- *E* is the degrees of freedom of the *error component and* should be somewhere between 10 and 20.

For example, if a study using laboratory animals is planned with four treatment groups (*T*=3), with eight animals per group, making 32 animals total (*N*=31), without any further stratification (*B*=0), then *E* would equal 28, which is above the cutoff of 20, indicating that sample size may be a bit too large, and six animals per group might be more appropriate.

23.5 QUALITATIVE RESEARCH

Qualitative research approaches sample size determination with a distinctive methodology that diverges from quantitative methods. Rather than relying on predetermined formulas or statistical calculations, it involves a subjective and iterative judgment throughout the research process In qualitative studies, researchers often adopt a subjective stance, making determinations as the study unfolds. Sample size determination in qualitative studies takes a different approach. It is generally a subjective judgment, taken as the research proceeds. One common approach is to continually include additional participants or materials until a point of "saturation" is reached. Saturation occurs when new participants or data cease to provide fresh insights, indicating that the study has adequately captured the diversity of perspectives or experiences within the chosen sample saturation is reached. The number needed to reach saturation has been investigated empirically.

Unlike quantitative research, qualitative studies face a scarcity of reliable guidance regarding sample size estimation prior to beginning the research. Imagine conducting in-depth interviews with cancer survivors, qualitative researchers may use data saturation to determine the appropriate sample size. If, over a number of interviews, no fresh themes or insights show up, saturation has been reached and more interviews might not add much to our knowledge of the survivor's experience. Thus, rather than following a preset statistical formula, the concept of attaining saturation serves as a dynamic guide for determining sample size in qualitative research.

There is a paucity of reliable guidance on estimating sample sizes before starting the research, with a range of suggestions given. In an effort to introduce some structure to the sample size determination process in qualitative research, a tool analogous to quantitative power calculations has been proposed. This tool, based on the negative binomial distribution, is particularly tailored for thematic analysis.

CHAPTER 24
OBSERVATIONAL STUDIES

In fields such as epidemiology, social sciences, psychology and statistics, an observational study draws inferences from a sample to a population where the independent variable is not under the control of the researcher because of ethical concerns or logistical constraints.

One common observational study is about the possible effect of a treatment on subjects, where the assignment of subjects into a treated group versus a control group is outside the control of the investigator. This is in contrast with experiments, such as randomized controlled trials, where each subject is randomly assigned to a treated group or a control group. Observational studies, for lacking an assignment mechanism, naturally present difficulties for inferential analysis.

24.1 MOTIVATION

The independent variable may be beyond the control of the investigator for a variety of reasons:

- A randomized experiment would violate ethical standards. Suppose one wanted to investigate the abortion – breast cancer hypothesis, which postulates a causal link between induced abortion and the incidence of breast cancer. In a hypothetical controlled experiment, one would start with a large subject pool of pregnant women and divide them randomly into a treatment group (receiving induced abortions) and a control group (not receiving abortions), and then conduct regular cancer screenings for women from both groups.
- Needless to say, such an experiment would run counter to common ethical principles. (It would also suffer from various confounds and sources of bias, e.g. it would be impossible to conduct it as a blind experiment.) The published studies investigating the abortion–breast cancer hypothesis generally start with a group of women who already have received abortions. Membership in this "treated" group is not controlled by the investigator: the group is formed after the "treatment" has been assigned.
- The investigator may simply lack the requisite influence. Suppose a scientist wants to study the public health effects of a community-wide

ban on smoking in public indoor areas. In a controlled experiment, the investigator would randomly pick a set of communities to be in the treatment group. However, it is typically up to each community and/or its legislature to enact a smoking ban.

- The investigator can be expected to lack the political power to cause precisely those communities in the randomly selected treatment group to pass a smoking ban. In an observational study, the investigator would typically start with a treatment group consisting of those communities where a smoking ban is already in effect.
- A randomized experiment may be impractical. Suppose a researcher wants to study the suspected link between a certain medication and a very rare group of symptoms arising as a side effect. Setting aside any ethical considerations, a randomized experiment would be impractical because of the rarity of the effect.
- There may not be a subject pool large enough for the symptoms to be observed in at least one treated subject. An observational study would typically start with a group of symptomatic subjects and work backwards to find those who were given the medication and later developed the symptoms. Thus a subset of the treated group was determined based on the presence of symptoms, instead of by random assignment.
- Many randomized controlled trials are not broadly representative of real-world patients and this may limit their external validity. Patients who are eligible for inclusion in a randomized controlled trial are usually younger, more likely to be male, healthier and more likely to be treated according to recommendations from guidelines. If and when the intervention is later added to routine-care, a large portion of the patients who will receive it may be old with many concomitant diseases and drug-therapies.

24.2 TYPES

- **Case-Control Study:** Study originally developed in epidemiology, in which two existing groups differing in outcome are identified and compared on the basis of some supposed causal attribute.
- **Cross-Sectional Study:** Involves data collection from a population, or a representative subset, at one specific point in time.
- **Longitudinal Study:** Correlational research study that involves repeated observations of the same variables over long periods of time. Cohort study and Panel study are particular forms of longitudinal study.

24.3 DEGREE OF USEFULNESS AND RELIABILITY

"Although observational studies cannot be used to make definitive statements of fact about the "safety, efficacy, or effectiveness" of a practice, they can:

- Provide information on 'real world' use and practice;
- Detect signals about the benefits and risks of...[the] use [of practices] in the general population;
- Help formulate hypotheses to be tested in subsequent experiments;
- Provide part of the community-level data needed to design more informative pragmatic clinical trials; and
- Inform clinical practice."

24.4 BIAS AND COMPENSATING METHODS

In all of those cases, if a randomized experiment cannot be carried out, the alternative line of investigation suffers from the problem that the decision of which subjects receive the treatment is not entirely random and thus is a potential source of bias. A major challenge in conducting observational studies is to draw inferences that are acceptably free from influences by overt biases, as well as to assess the influence of potential hidden biases. The following are a non-exhaustive set of problems especially common in observational studies.

24.4.1 Matching Techniques Bias

In lieu of experimental control, multivariate statistical techniques allow the approximation of experimental control with statistical control by using matching methods. Matching methods account for the influences of observed factors that might influence a cause-and-effect relationship. In healthcare and the social sciences, investigators may use matching to compare units that nonrandomly received the treatment and control. One common approach is to use propensity score matching in order to reduce confounding, although this has recently come under criticism for exacerbating the very problems it seeks to solve.

24.4.2 Multiple Comparison Bias

Multiple comparison bias can occur when several hypotheses are tested at the same time. As the number of recorded factors increases, the likelihood increases that at least one of the recorded factors will be highly correlated with the data output simply by chance.

24.4.3 Omitted Variable Bias

An observer of an uncontrolled experiment (or process) records potential factors and the data output: the goal is to determine the effects of the factors. Sometimes the recorded factors may not be directly causing the differences in the output. There may be more important factors which were not recorded but are, in fact, causal. Also, recorded or unrecorded factors may be correlated which may yield incorrect conclusions.

24.4.4 Selection Bias

Another difficulty with observational studies is that researchers may themselves be biased in their observational skills. This would allow for researchers to (either consciously or unconsciously) seek out the information they're looking for while conducting their research. For example, researchers may exaggerate the effect of one variable, or downplay the effect of another: researchers may even select in subjects that fit their conclusions. This selection bias can happen at any stage of the research process. This introduces bias into the data where certain variables are systematically incorrectly measured.

24.5 QUALITY

A 2014 (updated in 2024) Cochrane review concluded that observational studies produce results similar to those conducted as randomized controlled trials. The review reported little evidence for significant effect differences between observational studies and randomized controlled trials, regardless of design. Differences need to be evaluated by looking at population, comparator, heterogeneity, and outcomes.

CHAPTER 25
DATA PROCESSING

Data processing is the collection and manipulation of digital data to produce meaningful information. Data processing is a form of *information processing*, which is the modification (processing) of information in any manner detectable by an observer.

25.1 FUNCTIONS

Data processing may involve various processes, including:

- **Validation:** Ensuring that supplied data is correct and relevant.
- **Sorting:** "Arranging items in some sequence and/or in different sets."
- **Summarization (Statistical) or (Automatic):** Reducing detailed data to its main points.
- **Aggregation:** Combining multiple pieces of data.
- **Analysis:** The "collection, organization, analysis, interpretation and presentation of data."
- **Reporting:** List detail or summary data or computed information.
- **Classification:** Separation of data into various categories.

25.2 HISTORY

The United States Census Bureau history illustrates the evolution of data processing from manual through electronic procedures.

25.2.1 Manual Data Processing

Although widespread use of the term *data processing* dates only from the 1950s, data processing functions have been performed manually for millennia. For example, bookkeeping involves functions such as posting transactions and producing reports like the balance sheet and the cash flow statement. Completely manual methods were augmented by the application of mechanical or electronic calculators. A person whose job was to perform calculations manually or using a calculator was called a "computer."

The 1890 United States Census schedule was the first to gather data by individual rather than household. A number of questions could be answered by making a check in the appropriate box on the form. From 1850 to 1880 the Census Bureau employed "a system of tallying, which, by reason of the increasing number of combinations of classifications required, became increasingly complex. Only a limited number of combinations could be recorded in one tally, so it was necessary to handle the schedules 5 or 6 times, for as many independent tallies." "It took over 7 years to publish the results of the 1880 census" using manual processing methods.

25.2.2 Automatic Data Processing

The term *automatic data processing* was applied to operations performed by means of unit record equipment, such as Herman Hollerith's application of punched card equipment for the 1890 United States Census. "Using Hollerith's punchcard equipment, the Census Office was able to complete tabulating most of the 1890 census data in 2 to 3 years, compared with 7 to 8 years for the 1880 census. It is estimated that using Hollerith's system saved some $5 million in processing costs" in 1890 dollars even though there were twice as many questions as in 1880.

25.2.3 Computerized Data Processing

Computerized data processing, or electronic data processing represents a later development, with a computer used instead of several independent pieces of equipment. The Census Bureau first made limited use of electronic computers for the 1950 United States Census, using a UNIVAC I system, delivered in 1952.

25.2.4 Other Developments

The term *data processing* has mostly been subsumed by the more general term *information technology* (IT). The older term "data processing" is suggestive of older technologies. For example, in 1996 the *Data Processing Management Association* (DPMA) changed its name to the *Association of Information Technology Professionals.* Nevertheless, the terms are approximately synonymous.

25.3 APPLICATIONS

25.3.1 Commercial Data Processing

Commercial data processing involves a large volume of input data, relatively few computational operations, and a large volume of output. For example, an insurance company needs to keep records on tens or hundreds of thousands of policies, print and mail bills, and receive and post payments.

25.3.2 Data Analysis

In science and engineering, the terms *data processing* and *information systems* are considered too broad, and the term *data processing* is typically used for the initial stage followed by a data analysis in the second stage of the overall data handling. Data analysis uses specialized algorithms and statistical calculations that are less often observed in a typical general business environment. For data analysis, software suites like SPSS or SAS, or their free counterparts such as DAP, gretl, or PSPP are often used. These tools are usually helpful for processing various huge data sets, as they are able to handle enormous amount of statistical analysis.

25.4 SYSTEMS

A data processing system is a combination of machines, people, and processes that for a set of inputs produces a defined set of outputs. The inputs and outputs are interpreted as data, facts, information etc. depending on the interpreter's relation to the system. A term commonly used synonymously with *data or storage (codes) processing system* is *information system*. With regard particularly to electronic data processing, the corresponding concept is referred to as electronic data processing system.

25.4.1 Simple Example

A very simple example of a data processing system is the process of maintaining a check register. Transactions— checks and deposits— are recorded as they occur and the transactions are summarized to determine a current balance. Monthly the data recorded in the register is reconciled with a hopefully identical list of transactions processed by the bank.

A more sophisticated record keeping system might further identify the transactions— for example deposits by source or checks by type, such as charitable contributions. This information might be used to obtain information like the total of all contributions for the year.

The important thing about this example is that it is a *system*, in which, all transactions are recorded consistently, and the same method of bank reconciliation is used each time.

CHAPTER 26
PRESENTATION OF DATA

In the realm of research, business, and communication, the effective presentation of data is crucial for conveying information clearly and persuasively. Data presentation refers to the methods and techniques used to organize, summarize, and visually communicate data to facilitate understanding and decision-making.

26.1 UNDERSTANDING THE IMPORTANCE OF DATA PRESENTATION

Data presentation is essential for several reasons. First, it transforms raw data into meaningful information, allowing stakeholders to make informed decisions. In an era where data is abundant, the ability to present complex information succinctly and understandably is more important than ever. Moreover, effective data presentation enhances comprehension and retention. Well-organized and visually appealing data can capture attention, making it easier for audiences to grasp key insights.

This is particularly important in business settings, where stakeholders may need to make quick decisions based on presented information. Clear data presentation also fosters transparency and trust, as it allows for easy verification and interpretation of findings.

26.2 TECHNIQUES FOR PRESENTING DATA

Several techniques are available for presenting data effectively, ranging from textual descriptions to visual aids. The choice of method depends on the nature of the data, the audience, and the context in which the data will be presented.

26.2.1 Tables

Tables are a fundamental method for presenting data, allowing for the organization of numbers, text, or both into rows and columns. They are particularly useful for displaying large amounts of data in an easily accessible format. Tables enable readers to compare different variables, track changes over time, and extract specific information quickly.

When using tables, it is essential to include clear headings and labels for easy navigation. However, tables can become overwhelming if they contain too much information. To enhance readability, it is advisable to limit the amount of data displayed and highlight key figures. Proper formatting, such as alternating row colors or bolding significant data points, can further improve clarity.

26.2.2 Charts and Graphs

Visual representations of data, such as charts and graphs, are powerful tools for conveying information quickly and effectively. They allow audiences to identify trends, patterns, and relationships within the data at a glance. Common types of charts and graphs include:

- **Bar Charts:** Ideal for comparing quantities across different categories. They display data using rectangular bars, with the length of each bar representing the value it represents.
- **Line Graphs:** Useful for showing trends over time, line graphs connect data points with lines, allowing for easy visualization of changes and patterns.
- **Pie Charts:** Effective for illustrating proportions and percentages, pie charts divide a circle into slices, each representing a category's contribution to the whole.
- **Scatter Plots:** Used to show relationships between two variables, scatter plots display individual data points on a Cartesian plane, allowing for the identification of correlations.

When creating charts and graphs, it is important to select the type that best represents the data and to ensure that they are accurately labeled. Additionally, colors and design elements should be chosen thoughtfully to enhance clarity rather than distract from the information being presented.

26.2.3 Infographics

Infographics combine text, images, and data visualizations to convey complex information in a visually appealing and easily digestible format. They are particularly effective for summarizing research findings, illustrating processes, or highlighting key statistics. Infographics engage audiences by presenting information in a narrative style, making it more relatable and memorable. When designing an infographic, it is essential to maintain a clear structure. A well-organized layout, with a logical flow of information, helps guide the viewer's understanding. Using consistent color schemes, fonts, and icons can also enhance visual appeal and reinforce branding.

26.2.4 Dashboards

Dashboards are interactive data presentation tools that allow users to monitor key metrics and performance indicators in real-time. They typically combine multiple visualizations, such as charts, graphs, and tables, into a single interface. Dashboards are commonly used in business and healthcare settings, where stakeholders need to track performance, analyze trends, and make data-driven decisions quickly. The design of a dashboard should prioritize user experience. Information should be organized logically, with the most critical data prominently displayed. Interactive elements, such as filters and drill-down options, can enhance usability by allowing users to customize their views and access more detailed information as needed.

26.3 THE ROLE OF VISUALIZATION IN DATA PRESENTATION

Data visualization is a critical aspect of data presentation that enhances understanding and retention. It involves transforming numerical data into graphical representations, making it easier to identify trends, patterns, and outliers. Effective data visualization employs principles of design, such as contrast, color theory, and layout, to guide the viewer's attention and emphasize key messages.

26.3.1 Choosing the Right Visualization

Selecting the appropriate visualization method is crucial for conveying the intended message.

Different types of data may require different visualizations to communicate effectively. For instance, categorical data may be best represented using bar charts, while continuous data may be more effectively displayed using line graphs. Understanding the characteristics of the data and the story it tells is essential for making informed decisions about visualization methods.

26.3.2 Avoiding Misleading Visualizations

While visualizations can enhance comprehension, they can also mislead if not designed carefully. Common pitfalls include distorted scales, inappropriate chart types, and excessive embellishments. For example, using a 3D pie chart may obscure differences in proportions, while using a truncated bar graph can exaggerate differences between values. To avoid misleading interpretations, it is essential to adhere to ethical standards in data visualization and prioritize accuracy.

26.4 BEST PRACTICES FOR EFFECTIVE DATA PRESENTATION

To ensure that data presentation is clear, impactful, and persuasive, several best practices should be followed:

26.4.1 Know Your Audience

Understanding the audience is vital for tailoring the presentation of data to their needs and preferences. Consider their level of expertise, familiarity with the topic, and specific interests. This understanding will inform the choice of data presentation methods and the complexity of the information presented.

26.4.2 Keep It Simple

Simplicity is key when presenting data. Avoid cluttering presentations with excessive information, as this can overwhelm the audience and obscure key messages. Focus on the most important findings and present them clearly and concisely.

26.4.3 Highlight Key Insights

Use visual elements such as color, bolding, and annotations to highlight key insights and findings. This helps draw the audience's attention to the most important aspects of the data, making it easier for them to grasp the essential messages.

26.4.4 Provide Context

Data presentation should provide context to help the audience understand the significance of the findings. Include relevant background information, comparisons to previous studies, or benchmarks to enhance the audience's comprehension of the data.

26.4.5 Practice Good Design Principles

Adhering to design principles, such as consistency, alignment, and contrast, can significantly enhance the effectiveness of data presentation. Use a consistent color palette, font style, and layout throughout the presentation to create a cohesive look. Ensure that visual elements are aligned properly and that there is sufficient contrast between text and background colors for readability.

26.5 CONCLUSION

The presentation of data plays a pivotal role in effectively communicating information, insights, and findings across various fields. By utilizing appropriate techniques, such as tables, charts, infographics, and dashboards, stakeholders can convey complex information clearly and persuasively. Effective data presentation enhances comprehension, supports decision-making, and fosters trust and transparency. As data continues to grow in importance, mastering the art of data presentation will be essential for professionals in research, business, and communication to navigate the complexities of the information landscape successfully.

BIBLIOGRAPHY

An Introduction to Statistical Methods and Data Analysis Author: R. Lyman Ott, et al. Publisher: Cengage Learning Year: 2010

Applied Biostatistics for the Health Sciences Author: Richard J. Rossi Publisher: Wiley Year: 2002

Applied Multivariate Statistical Analysis Author: Richard A. Johnson, et al. Publisher: Prentice Hall Year: 2002

Applied Regression Analysis and Generalized Linear Models Author: John Fox Publisher: SAGE Publications Year: 2016

Biostatistics for Clinical and Public Health Research Author: Michael T. Long Publisher: Jones & Bartlett Learning Year: 2018

Biostatistics: A Foundation for Analysis in the Health Sciences Author: Daniel W. McNeil, et al. Publisher: Wiley Year: 2010

Biostatistics: A Guide to Design, Analysis, and Discovery Author: J. Richard L. R. C. McCulloch Publisher: Wiley Year: 2005

Biostatistics: A Methodology for the Health Sciences Author: Gerald van Belle, et al. Publisher: Wiley Year: 2004

Biostatistics: An Introduction Using R Author: Babe F. H. Szafran Publisher: John Wiley & Sons Year: 2016

Biostatistics: An Introductory Text Author: B. M. A. van der Meer Publisher: Kluwer Academic Publishers Year: 1997

Clinical Trials: A Practical Guide Author: Duolao Wang, et al. Publisher: Wiley Year: 2016

Designing Clinical Research Author: Stephen B. Hulley, et al. Publisher: Lippincott Williams & Wilkins Year: 2013

Epidemiology: Beyond the Basics Author: Moyses Szklo, et al. Publisher: Jones & Bartlett Learning Year: 2014

Fundamentals of Biostatistics Author: Bernard Rosner Publisher: Brooks/Cole Year: 2011

Introduction to Biostatistics Author: Robert R. Sokal, et al. Publisher: W. H. Freeman Year: 1995

Introduction to the Practice of Statistics Author: David S. Moore, et al. Publisher: W. H. Freeman Year: 2016

Introductory Statistics Author: Prem S. Mann Publisher: Wiley Year: 2010

Modern Epidemiology Author: Kenneth J. Rothman, et al. Publisher: Lippincott Williams & Wilkins Year: 2008

Practical Statistics for Medical Research Author: Douglas G. Altman Publisher: Chapman and Hall Year: 1991

Principles of Biostatistics Author: Marcello Pagano, et al. Publisher: Duxbury Press Year: 2000

Research Design and Statistical Analysis Author: Arnold A. Lazarsfeld, et al. Publisher: Lawrence Erlbaum Associates Year: 1999

Research Methods in Psychology: Evaluating a World of Information Author: Beth Morling Publisher: W. W. Norton & Company Year: 2017

Statistical Analysis of Epidemiologic Data Author: Richard H. Smith Publisher: Academic Press Year: 2010

Statistical Analysis with Missing Data Author: Donald B. Rubin Publisher: Wiley Year: 1987

Statistical Methods for Rates and Proportions Author: Joseph L. Fleiss Publisher: Wiley Year: 2003

Statistical Methods for the Behavioral Sciences Author: Gregory J. Privitera Publisher: SAGE Publications Year: 2017

Statistical Methods for the Health Sciences Author: Daniel W. McNeil, et al. Publisher: Wiley Year: 2006

Statistical Methods in Epidemiology Author: Ray M. Merrill Publisher: Jones & Bartlett Learning Year: 2013

Statistical Methods in Health Care Research Author: Barbara Hazard Munro Publisher: Lippincott Williams & Wilkins Year: 2005

Statistical Techniques in Business and Economics Author: Douglas A. Lind, et al. Publisher: McGraw-Hill Year: 2018

Statistics for Business and Economics Author: Paul Newbold, et al. Publisher: Pearson Year: 2019

Statistics for Epidemiology and Public Health Author: Ian R. S. M. D. MacMahon Publisher: Wiley Year: 2014

Statistics for Health Care Professionals Author: Aviva Petrie, et al. Publisher: SAGE Publications Year: 2016

INDEX

www.ingramcontent.com/pod-product-compliance
Lightning Source LLC
LaVergne TN
LVHW062119200926
848484LV00007B/1716
9781787154407